Reviews of

115 Physiology Biochemistry and Pharmacology

Editors
M. P. Blaustein, Baltimore · O. Creutzfeldt, Göttingen
H. Grunicke, Innsbruck · E. Habermann, Gießen
H. Neurath, Seattle · S. Numa, Kyoto
D. Pette, Konstanz · B. Sakmann, Heidelberg
M. Schweiger, Innsbruck · U. Trendelenburg, Würzburg
K. J. Ullrich, Frankfurt/M · E. M. Wright, Los Angeles

With 24 Figures and 4 Tables

Springer-Verlag Berlin Heidelberg GmbH

ISBN 978-3-662-41743-0   ISBN 978-3-662-41884-0 (eBook)
DOI 10.1007/978-3-662-41884-0

Library of Congress-Catalog-Card Number 74-3674

© Springer-Verlag Berlin Heidelberg 1990
Originally published by Springer-Verlag Berlin Heidelberg New York 1990
Softcover reprint of the hardcover 1st edition 1990

Typesetting: K + V Fotosatz GmbH, Beerfelden
2127/3130-543210 – Printed on acid-free paper

Contents

Indexed in Current Contents

Rev. Physiol. Biochem. Pharmacol., Vol. 115
© Springer-Verlag 1990

Diversity of Sodium Channels in Adult and Cultured Cells, in Oocytes and in Lipid Bilayers

BERTHOLD NEUMCKE

Contents

1 Introduction

Ionic channels in biological membranes enable the passive movement of ions between the extra- and intracellular solutions and are generally named after the main permeant ion. Thus open sodium channels are selective for Na$^+$ ions and have only minor permeability for other cations and anions. Such

I. Physiologisches Institut, Universität des Saarlandes, D-6650 Homburg (Saar), FRG

Na$^+$-preferring channels are present in electrically excitable membranes of nerve and muscle and also in many inexcitable cells, e.g. epithelial cells (Fuchs et al. 1977; Palmer 1987) and light receptor cells of vertebrate eyes (Yau and Nakatani 1984; Hodgkin et al. 1985). However, this review and the following chapters in this volume are devoted exclusively to voltage-gated channels.

At first sight all sodium channels of neural and muscular origin appear to be quite homogeneous. They open transiently upon depolarization and thus produce the characteristic fast activation and slower inactivation phases of the macroscopic sodium currents (compare Figs. 2, 4, 7a, b, and Hille 1984 p. 73, Fig. 9). The first indication of channel heterogeneity was the discovery of distinct toxin effects on sodium channels in various excitable cells. Thus, tetrodotoxin blocks neural sodium channels even in the nM concentration range (Narahashi et al. 1964; Nakamura et al. 1965), while a concentration of several μM is necessary for an equally effective block of sodium channels in heart muscle (Brown et al. 1981; Cohen et al. 1981). Moreover, embryonic neurons and muscles often have tetrodotoxin-resistant sodium channels (Spitzer 1979; Frelin et al. 1984). Differences between various sodium channels are also revealed by the use of toxin ATX II from the sea anemone, *Anemonia sulcata*. This toxin slows the sodium inactivation in myelinated nerves at a concentration of several μM (Bergman et al. 1976; Ulbricht and Schmidtmayer 1981), but only a few nM prolong action potentials and modify inactivation of sodium currents in heart muscle (Ravens 1976; Isenberg and Ravens 1984). In addition, sodium channels in the surface membrane and transverse tubules of frog skeletal muscle can be distinguished by their different binding affinities and action kinetics with tetrodotoxin, ethylenediamine derivatives of tetrodotoxin, and *Centruroides* or *Tityus* scorpion toxins (Jaimovich et al. 1982, 1983; Arispe et al. 1988). Finally, conotoxins from the snail *Conus geographus* can be used to discriminate between muscle, cardiac and neuronal sodium channels (Moczydlowski et al. 1986; Kobayashi et al. 1986) and in addition between two channel subtypes in rat muscle (Gonoi et al. 1987).

Recent determinations of the amino acid sequence of sodium channel proteins have corroborated the channel diversity. Despite a striking degree of homology, minor differences have been found between channels in the electroplax of *Electrophorus electricus* (Noda et al. 1984), in rat brain (Noda et al. 1986a) and in *Drosophila* (Salkoff et al. 1987). Furthermore, three sodium channel subtypes have been identified in rat (Kayano et al. 1988) which are distributed differently in skeletal muscle (Haimovich et al. 1987) and in brain (Gordon et al. 1987). At present the physiological significance of the various sodium channels and their preferred locations in a variety of tissues is unknown (Barchi 1987). Also, the structural changes in the channel molecules underlying the gating reactions have not yet been identified, though first reports have appeared on the participation of specific regions or single amino acids of the channel protein in the formation of the membrane pore (Oiki et al. 1988), in the activation of the channel (Stühmer et al. 1989), and in the process of sodium inactivation (Meiri et al. 1987; Vassilev et al. 1988; Krafte et al. 1988; Stühmer et al. 1989) and in the binding of α-scorpion toxins (Tejedor and Catterall 1988). Therefore, it cannot yet be decided whether the different electrophysiological properties of various sodium channels are due to (a) variations in the amino acid sequence of the channel protein, (b) the participation of smaller peptides which copurify with the main α peptide in mammalian preparations (Barchi 1983; Hartshorne and Catterall 1984), (c) varying degrees of post-translational glycosylation of the channel protein or (d) different lipid domains around the sodium channel in the excitable membrane. The following discussion of the diversity of sodium channels in various excitable cells is, therefore, mainly descriptive and interpretations in terms of channel structures or lipid environments are necessarily rather speculative. In particular, differences in the electrophysiological properties of the channels derived from recent measurements of macroscopic sodium currents and gating currents in multichannel preparations and from single channel currents are reviewed. A problem with this comparison is the difficulty of distinguishing between variations of real channel properties and apparent differences which are due to the experimental protocol or to inadequate consideration of artifacts. For example, large sodium currents from many sodium channels may be distorted by the voltage drop across the resistance in series with the membrane being studied (Hodgkin et al. 1952) producing a voltage shift of gating parameters. Similar voltage shifts in patch clamp experiments on sodium channels, seen when cell-attached or cell-excised patches from the same cells are compared, are still unexplained (Fenwick et al. 1982; Nagy et al. 1983; Cachelin et al. 1983). Despite these ambiguities several real differences exist between the electrophysiological properties of sodium channels in adult or cultured cells and those of channels expressed in oocytes or incorporated into artificial lipid bilayers. A critical discussion of these differences is presented in this review with the hope that an interpretation at the molecular level will be possible in the near future.

2 Macroscopic Sodium Currents

The classical description of ionic currents in the squid giant axon by Hodgkin and Huxley (1952) is also widely used for other electrically excitable membranes. In the Hodgkin-Huxley equations for the sodium current the term m^3h can be considered as the probability of an open sodium channel. The symbol m denotes the activation variable, h the inactivation variable, and the kinetics of m and h are described by first-order differential equations. Thus, the current relaxation at a given membrane potential is specified by the parameters m_∞ (steady state activation variable), h_∞ (steady state inactivation variable), τ_m (time constant of sodium activation) and τ_h (time constant of sodium inactivation). The steady state properties of sodium activation and inactivation in various excitable cells are discussed in Sect. 2.1 and the kinetics of these gating processes are discussed in Sects. 2.3, 2.4.

2.1 Steady State Sodium Activation and Inactivation

Figure 1 compares the voltage dependencies of the quantities h_∞, m_∞ and m_∞^3 in squid giant axons (Fig. 1a), in mouse myocardial cells (Fig. 1b), in Ranvier nodes of rat sciatic nerve (Fig. 1c), and in *Xenopus* oocytes with

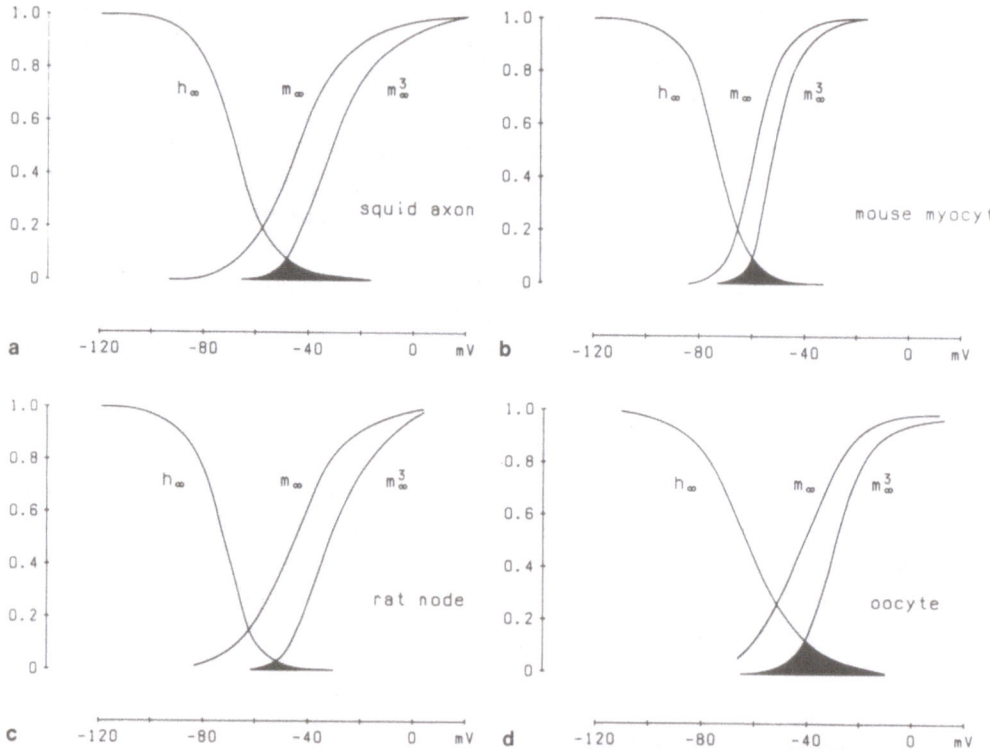

Fig. 1a–d. Steady state parameters m_∞, m_∞^3 and h_∞ of sodium activation and inactivation in a variety of excitable cells. *Filled areas* represent the overlap of the m_∞^3 and h_∞ curves. The membranes used are **a** squid giant axon with an assumed resting potential of -70 mV, **b** single myocardial mouse cell, **c** rat sciatic node of Ranvier with an assumed resting potential of -78 mV, and **d** rat brain sodium channels type II expressed in *Xenopus* oocytes. (**a** from Hodgkin and Huxley 1952; **b** from Benndorf et al. 1985; **c** from Neumcke et al. 1987; **d** from Stühmer et al. 1987)

implanted sodium channels of type II from rat brain (Fig. 1 d). The parameters are plotted as a function of the membrane potential, though the exact value of the resting potential is unknown or not given for several of the preparations selected for Fig. 1; therefore, the location of the midpoint potentials could be affected by this uncertainty. However, the overlap between the activation and inactivation curves is independent of the assumed resting potential. The "window" between the h_∞ and m_∞^3 curves (shaded areas in Fig. 1) is of particular interest because it reflects the potential range in which steady state sodium currents are observed. The amplitude of these stationary currents is proportional to $m_\infty^3 \cdot h_\infty$ and thus varies with the size of the window. Figure 1 reveals clear differences between the stationary sodium currents in various preparations. These differences cannot be attributed to different species because the largest deviations are found for rat sodium channels present either in nodes of Ranvier (Fig. 1 c) or expressed in *Xenopus* oocytes (Fig. 1 d). Possible explanations for this are that the channel types present in rat brain are different from those in rat peripheral nerve (Gordon et al. 1987) or that rat sodium channels are modulated by the foreign environment of the oocyte membrane.

In the examples illustrated in Fig. 1, the different sizes of the windows between the h_∞ and m_∞^3 curves are mainly due to variations in the inactivation variable h_∞, while the potential dependence of the activation variable m_∞ is rather uniform for different sodium channels. For example, the m_∞ curves for rat sodium channels in nodes (Fig. 1 c) and in oocyte membranes (Fig. 1 d) are almost identical, whereas the h_∞ curve for the oocyte channel is flatter and shifted towards more positive potentials than that for the peripheral nerve sodium channel (Stühmer et al. 1987). In general, the midpoint potentials E_h of h_∞ curves show great variability between different cells. The largest negative E_h value, of -109 mV, has been reported for mouse pancreatic B cells (Plant 1988), whereas an E_h of -35 mV reported for bovine chromaffin cells (Fenwick et al. 1982) is an extremely small negative midpoint potential. However, varying overlaps between inactivation and activation curves may also be due to voltage shifts of the m_∞ parameter, e.g. in mammalian cultured Schwann cells (Shrager et al. 1985) and frog taste receptor cells (Avenet and Lindemann 1987) the activation is shifted to more positive potentials, whereas inactivation has a normal voltage dependence.

Sodium channels with different m_∞ or h_∞ midpoint potentials are not always confined to different cells, but they may also coexist in the same preparation, e.g. adult (tetrodotoxin-sensitive) and juvenile (tetrodotoxin-insensitive) sodium channels with different h_∞ curves are present in myoballs cultivated from rat (Ruppersberg et al. 1987) or human (Pröbstle et al. 1988) skeletal muscle.

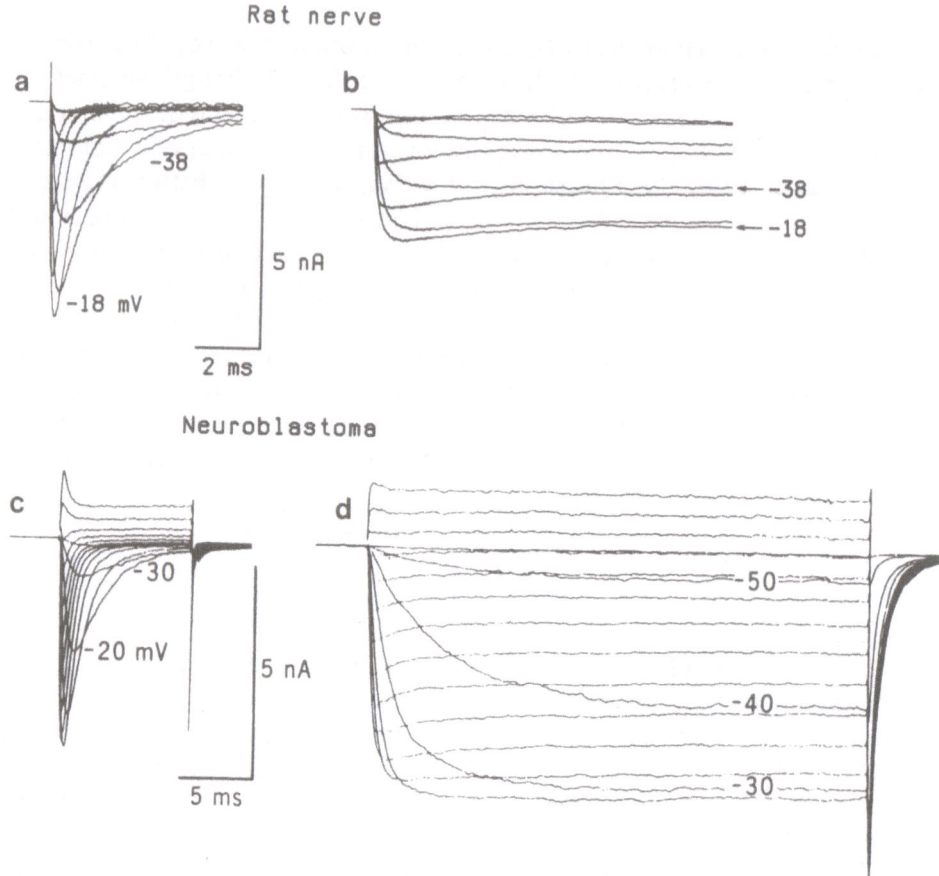

Fig. 2a – d. Effects of inactivation-modifying substances on sodium channels in a node of Ranvier of rat sciatic nerve (**a, b**) and in an N18 neuroblastoma cell (**c, d**). **a** Control sodium currents in rat nerve at test potentials of −54 (slowest current activation), −46, −38, −30, −18, 2, 22, and 42 mV (fastest current activation). **b** Sodium currents 2 min after extracellular application of 0.5 mM chloramine-T at the same test potentials as in **a. c** Control sodium currents in a neuroblastoma cell at test potentials from −60 to 90 mV in intervals of 10 mV. **d** Sodium currents 60 min after intracellular application of 0.25 mg/ml papain. (**a, b** from Neumcke et al. 1987, with an assumed resting potential of −78 mV; **c, d** from Gonoi and Hille 1987)

2.2 Shift of m_∞ Curve by Inactivation Modifiers

In the Hodgkin-Huxley equations the probability of an open sodium channel (m^3h) is formulated by the product of the probabilities of a channel being activated (m^3) and not being inactivated (h). Hence, it is assumed that the two gating processes of sodium activation and inactivation proceed independently of each other. This concept is supported by numerous pharmacological experiments in which strong modification of sodium inactivation had no or only little effect on sodium activation (Hille 1984). As an example, Fig. 2a, b

Fig. 3a, b. Normalized peak sodium conductances before and after modification of sodium inactivation in a node of rat sciatic nerve (**a**) and in an N18 neuroblastoma cell (**b**). Values in **a** before (○) and after (+) application of extracellular chloramine-T were obtained from the experiment illustrated in Fig. 2a, b. Values in **b** before (○) and after (●) inactivation is modified by intracellularly applied papain are from Gonoi and Hille (1987)

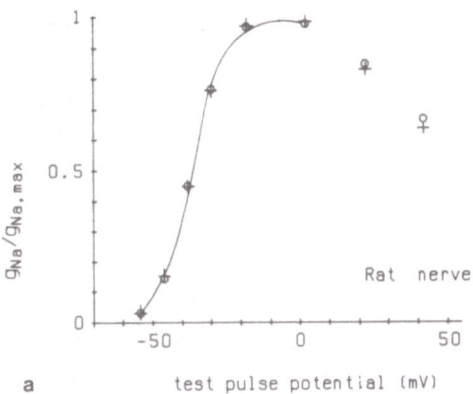

a

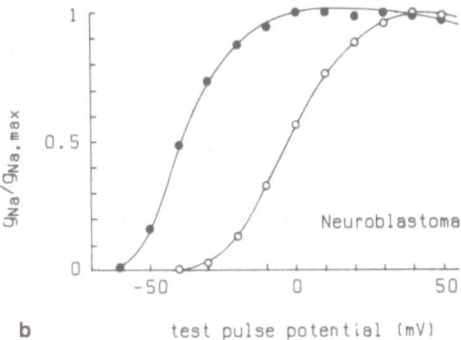

b

shows an experiment on a myelinated fibre of rat sciatic nerve and depicts sodium currents at various depolarizations under control conditions (Fig. 2a) and after addition of chloramine-T to the extracellular solution (Fig. 2b). Chloramine-T reduces sodium currents in myelinated nerve at all test potentials, induces a marked slowing of the process of sodium inactivation, but has little effect on the kinetics and voltage dependence of sodium activation (Wang 1984; Neumcke et al. 1987). Different results are obtained using N18 neuroblastoma cells (Gonoi and Hille 1987): As illustrated in Fig. 2c, d, inactivation in this cultured cell is almost completely destructed by the action of intracellular papain, while sodium currents at all test potentials are *increased*, and in addition this treatment produces a strong alteration in sodium activation. For example, a test pulse of −40 mV elicits only a small sodium current under control conditions (Fig. 2c) but a significant sustained current after channel modification (Fig. 2d).

The different effects of inactivation-modifying substances on peak sodium currents in nerve and neuroblastoma cells are clearly seen in Fig. 3, in which normalized peak sodium conductances g_{Na} have been plotted as a function of the test potential. With rat nerve (Fig. 3a) the conductances are almost identical before and after modification (the decline of g_{Na} at large depolarizations is due to the use of sodium conductances instead of permeabilities (Dodge and Frankenhaeuser 1959)). On the other hand, there is a $-24\,mV$ shift of the conductance curve with the neuroblastoma cell (Fig. 3b), showing that sodium channels become activated at more negative potentials after modification of sodium inactivation. The corresponding large negative voltage shift of the m_∞ curve produced by inactivation-modifying substances seems to be a genuine property of the neuroblastoma cell (Gonoi and Hille 1987), because no or much smaller shifts are observed in sodium channels of other preparations. Thus no voltage shift of activation parameters occurs in rat nerve after treatment with chloramine-T (Fig. 3a) or in squid giant axons after destructing sodium inactivation by internal pronase (Armstrong et al. 1973). Small negative m_∞ shifts of some mV have been reported for frog nerve after treatment with scorpion venom (Koppenhöfer and Schmidt 1968) or with toxin II of *Anemonia sulcata* (Neumcke et al. 1985) and for squid giant axons after administration of chloramine-T (Wang et al. 1985). Hence, sodium channels in peripheral nerves show no or only small changes in the process of sodium activation after modification of sodium inactivation. Similarly, various agents which slow inactivation in frog skeletal muscle have only minor effects on the activation mechanism (Nonner et al. 1980). The large negative m_∞ shifts observed by Gonoi and Hille (1987) in N18 neuroblastoma cells are not typical of other cultured cells: for example, Patlak and Horn (1982) did not find a change in sodium activation in rat myotubes after administration of N-bromoacetamide, while Vandenberg and Horn (1984) described a slight *positive* voltage shift of the peak sodium conductance curve in trypsin-treated GH_3 cells. According to the Hodgkin-Huxley formalism, modification of the h inactivation process should not change the m activation parameters, and this seems to be approximately fulfilled for most of the sodium channels. However, the large m_∞ shifts observed in neuroblastoma cells indicate that the independence between the m- and h-gating processes inherent in the Hodgkin-Huxley scheme is no longer valid. Instead, a reaction diagram, in which the processes of sodium activation and inactivation are coupled to each other, seems to be more appropriate to describe voltage shifts of the m_∞ curve produced by inactivation modifying substances (Gonoi and Hille 1987; see also Sect. 5.1).

2.3 Kinetics of Sodium Activation

Upon depolarization the increase in the sodium conductance follows an S-shaped time course, whereas on repolarization the decrease is exponential.

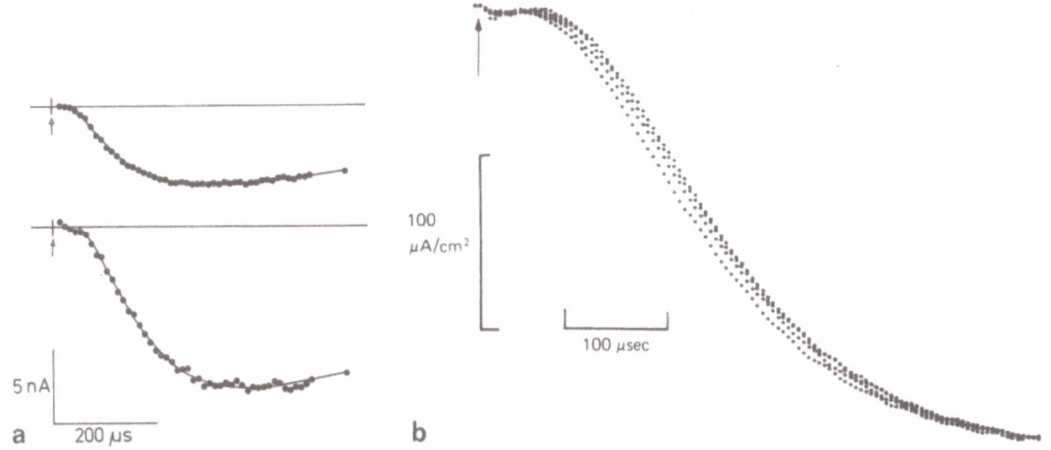

Fig. 4a, b. Effect of prepulses on the rise of sodium current in a node of frog sciatic nerve (**a**) and in squid giant axon (**b**). The onset of the test pulses is marked with *arrows*. **a** 47-ms prepulses to −66 mV (*top*) and −114 mV (*bottom*), with test pulses to −6 mV, temperature 10 °C. **b** 100-ms prepulses to −90 (*extreme left-hand trace*), −120, −150, and −180 mV (*extreme right-hand trace*), with test pulses to 20 mV, temperature 7 °C (**a** from Neumcke et al. 1976b, with an assumed resting potential of −70 mV; **b** from Keynes and Kimura 1983)

This behaviour can be described by raising the activation variable m to the third power (Hodgkin and Huxley 1952). The m^3 formulation of sodium activation in the squid giant axon could also be applied to sodium channels in other excitable membranes, but sometimes slight modifications have been suggested. For example, in myelinated nerve fibres of *Xenopus laevis* the expression m^2 describes the rise of the sodium current somewhat better than m^3 (Frankenhaeuser 1960). The difference has been interpreted by different numbers of *m* gating particles controlling the activation of a single sodium channel: three particles in the squid giant axon and only two at the node of Ranvier (Keynes and Rojas 1976).

If the number of *m* gating particles is integer, however, it is difficult to reconcile this number with the graded effects of prepulses on the initial rise in the sodium conductance (Fig. 4). Making the prepulses more negative delays sodium activation in the squid giant axon (experiments by H.M. Fishman, see Fig. 12 of Offner 1972; Armstrong and Bezanilla 1974; Keynes and Rojas 1976; Keynes and Kimura 1983; Taylor and Bezanilla 1983), in frog myelinated nerve (Neumcke et al. 1976b), and in frog skeletal muscle (Collins et al. 1982a). In order to account for the prepulse effects on sodium activation noninteger and voltage-dependent exponents *a* are required in the m^a term. Hence, the parameter *a* can no longer be interpreted as the number of gating particles in a sodium channel.

The range of the *m* exponent shows remarkable differences between the squid giant axon and the frog node of Ranvier. In the giant axon there are

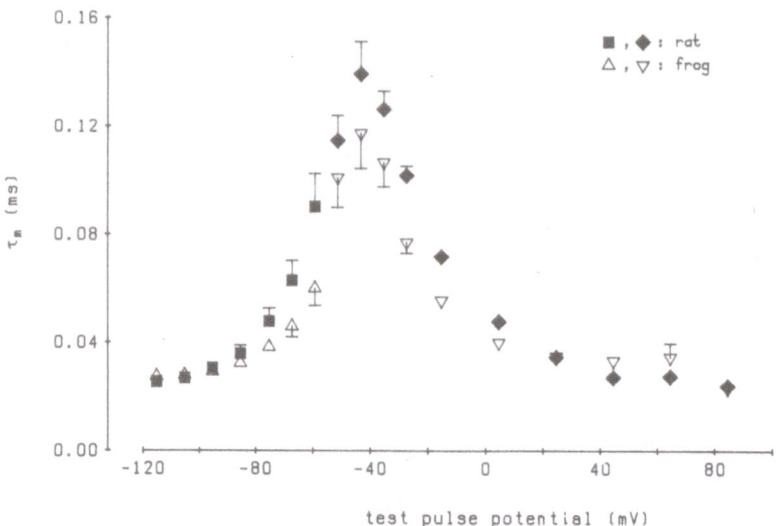

Fig. 5. Time constants τ_m of sodium activation in rat nodes (*filled symbols*) and frog sciatic nerve (*open symbols*). Values at test potentials $\geqslant -50$ mV ($\blacklozenge$, $\triangledown$) were obtained from the activation of sodium currents at these potentials, and values at test potentials $\leqslant -58$ mV ($\blacksquare$, $\triangle$) were obtained from sodium tail currents at these potentials following 0.3-ms prepulses to -14 mV. Temperature 20 °C; assumed resting potential of -74 mV for rat and frog nerve fibres. (From Neumcke et al. 1987)

only small variations of the exponent from axon to axon, i.e. between 2.9 and 4.4, and no significant variations with pulse potentials have been observed (Keynes and Kimura 1983). On the other hand, the value of a must be between 1.6 and 8.1 in order to fit the rise of sodium currents in frog nerve fibres at various prepotentials and test potentials (Neumcke et al. 1976b).

There are great variations between preparations in the time constants of sodium activation, even when the rise of the sodium currents is fitted with the same equations and the experiments are performed at the same temperature. An extreme case was reported by Sah et al. (1988) who found that in some of the cells dissociated from guinea pig hippocampal slices the kinetics of the sodium currents are by one order of magnitude faster than in the majority of the investigated cells. The most likely interpretation that has been suggested is that different types of sodium channels with slow and fast kinetics exist in different cells of the hippocampus. Different activation time constants were also found for sodium channels in frog myelinated nerve and frog skeletal muscle; for this species the activation process in nerve is five times faster than in muscle (Campbell and Hille 1976). Comparatively small differences in the τ_m values have been described for frog and rat nerve fibres (Neumcke et al. 1987). Figure 5 was reproduced from this study and shows that the τ_m values in the rat are 15% – 50% larger than those in the frog at moderate depolarizations. This does not necessarily imply the presence of different types of sodium channels in amphibian and mammalian nerve, because the relatively small differences in τ_m could also arise from different lipid environments of the channels.

2.4 Kinetics of Sodium Inactivation

As pointed out in Sect. 2.2, the process of sodium inactivation may be considered to be independent of or coupled to the reactions activating the sodium channels. Although these two views imply totally different mechanisms of channel inactivation, discrimination between the alternatives is difficult to achieve from an analysis of macroscopic sodium currents. For example, a positive voltage shift of the h_∞ curve with larger test pulses should be indicative of coupling between activation and inactivation (Hoyt 1963, 1968), but the h_∞ shifts observed in *Myxicola* giant axons could not be distinguished reliably from artifacts produced by the resistance in series with the axon membrane (Goldman and Schauf 1972). Another possible method of discriminating between reaction schemes with independent or coupled inactivation is to study the development of sodium inactivation. While an independent inactivation reaction would start immediately at the moment of depolarization, a delay should exist in coupled schemes due to the preceding activation. Indeed, a delay in the development of sodium inactivation has been observed in *Myxicola* giant axons (deviations between measured sodium currents and interrupted curve in Fig. 6a) and taken as evidence for activation-inactivation coupling (Goldman and Schauf 1972; Goldman and Kenyon 1982). However, as pointed out by Hille (1976), the delay "shows primarily that inactivation is not a simple first-order process and does not go far towards demonstrating any causal relation to the degree of activation". In squid giant axons the time course of inactivation is nearly exponential (Fig. 6b) and a possible delay in this gating reaction is smaller than 100 µs (Gillespie and Meves 1980). Similarly, there is no indication of a delay in the development of sodium inactivation in myelinated nerve fibres of *Xenopus laevis* (Kniffki et al. 1981). The crayfish giant axon seems to represent an intermediate case between the two extremes. Bean (1981) reported that there is a delay in sodium inactivation in this preparation, but less than would be expected from the activation time. Thus, inactivation in crayfish axons is only a little delayed and can occur already before the opening of sodium channels. It has been concluded from single channel experiments that a similar behaviour seems likely in sodium channels of rat myotubes and neuroblastoma cells (Horn et al. 1981a; Aldrich and Stevens 1983).

A distinction between various sodium channels may also be made by comparing the inactivation time constants which have been derived from one and two pulse experiments. In the Hodgkin-Huxley formalism the inactivation

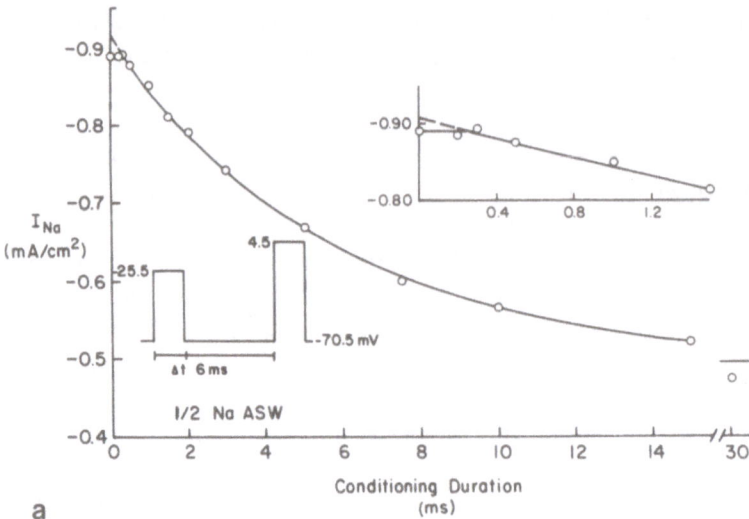

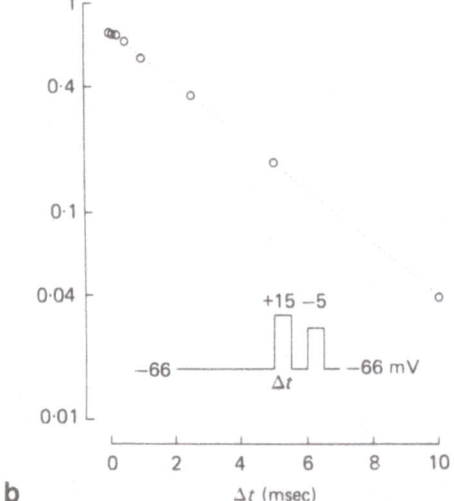

Fig. 6a, b. Development of sodium inactivation in *Myxicola* (**a**) and squid (**b**) giant axons. Peak sodium currents I_{Na} (**a**) and normalized peak sodium currents (**b**) are plotted as functions of the duration Δt of conditioning depolarizing prepulses (see pulse pattern in *insets*). Exponential decays of sodium currents are indicated by the *interrupted* and *full curve* in (**a**) and by the *dotted line* in (**b**). The *top inset* in **a** shows the initial part of the curve and an inactivation delay of 285 µs on an expanded time scale. Note linear (**a**) and logarithmic (**b**) scaling of ordinate. (**a** from Goldman and Kenyon 1982; **b** from Gillespie and Meves 1980)

variable h is described by a first-order differential equation; consequently, the inactivation time constant depends only on the test potential. On the other hand, if inactivation is not a first-order process, or if activation-inactivation coupling is considered, different time constants should follow from the decline of sodium currents (time constant τ_h) and from the recovery or development of sodium inactivation following conditioning prepotentials (time constant τ_c). Such a difference between the τ_h and τ_c values has been reported for the giant axons of *Myxicola* (Goldman and Schauf 1973) and of lobster (Oxford and Pooler 1975), whereas practically the same τ_h and τ_c time values have been found for squid (Bezanilla and Armstrong 1977; Gillespie and

Meves 1980), crayfish (Bean 1981), and myelinated nerve fibres of frog (Chiu 1977) and rat (Neumcke et al. 1987).

Large deviations between the inactivation time constants similar to the great variations in the activation kinetics are also seen in different cells (see Sect. 2.3). For example, at 0 mV and 8 °C the τ_h values of squid and crayfish giant axons are 2.5 and 0.45 ms, respectively (Swenson 1980). A somewhat smaller variation between different preparations is observed when the inactivation time constant τ_h is referred to the activation time constant τ_m or to the time t_p, the peak of the sodium current. τ_h/t_p values have been compiled for various excitable cells (Carbone and Lux 1986, Table 3). The extreme ratios are $\tau_h/t_p = 1$ for mouse neuroblastoma cells (Moolenaar and Spector 1978) and $\tau_h/t_p = 3.6$ for *Aplysia* neurons (Adams and Gage 1979). These examples illustrate that the activation and inactivation rates of various sodium channels are not linked to each other. Instead, the two gating processes may undergo separate modifications in different excitable cells.

In the Hodgkin-Huxley equations the development and the removal of sodium inactivation are described by a single time constant τ_h in the millisecond time range. Frequently, additional slow and ultraslow inactivation processes are induced in sodium channels by longer conditioning prepulses. Such slow reactions, some of which are dependent on the extracellular K^+ concentration, have been reported for a great variety of preparations, e.g. giant axons of lobster (Narahashi 1964), squid (Adelman and Palti 1969; Chandler and Meves 1970b) and *Myxicola* (Schauf et al. 1976); frog myelinated nerve (Peganov et al. 1973; Fox 1976; Brismar 1977); frog, rat and human skeletal muscle (Collins et al. 1982a; Almers et al. 1983, 1984; Simoncini and Stühmer 1987; Ruff et al. 1987); rabbit cardiac Purkinje fibres (Carmeliet 1987); and neuroblastoma cells (Gonoi and Hille 1987). In order to interpret this common phenomenon it has been suggested that long-lasting hyperpolarizations increase the number of available sodium channels (Neumcke et al. 1976a). Indeed, a fluctuation analysis of sodium currents in frog nerve revealed a higher number of sodium channels per node at more negative holding potentials (Neumcke and Stämpfli 1983).

The time constants of slow sodium inactivation normally are in the range between 100 ms and several minutes, and the processes are studied by recording peak sodium currents as a function of the duration of conditioning prepulses. Exceptions to this are two distinct inactivation reactions which are in the millisecond time range in myelinated nerve fibres (Chiu 1977; Nonner 1980; Neumcke and Stämpfli 1982) and in cardiac muscle (Benndorf and Nilius 1987; Clark and Giles 1987; Antoni et al. 1988), and which produce a biphasic decay of sodium currents during a single depolarization (see Fig. 7b). In neural membranes this behaviour seems to be more pronounced for myelinated nerve than for other axon preparations, e.g. Fig. 7a illustrates that the decline of sodium currents in squid giant axon follows a single exponen-

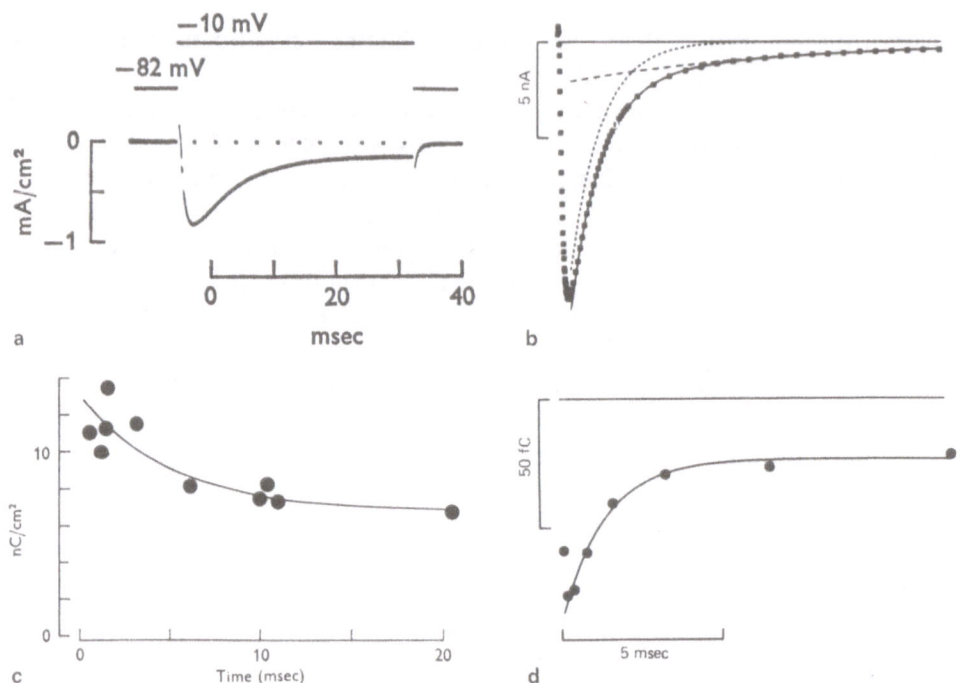

Fig. 7a–d. Sodium inactivation (**a, b**) and charge immobilization (**c, d**) in squid giant axon (**a, c**) and in frog myelinated nerve (**b, d**). **a** Exponential decline of sodium current at −10 mV up to 40 ms, τ_h = 7.75 ms, temperature 0 °C; **b** Biphasic decay of sodium current at −6 mV with time constants of 0.94 and 6.7 ms, temperature 13 °C, the time scale is the same as in **d**; **c, d** Fast off-charge displacements upon repolarization as functions of the duration of test pulses, plotted upward (**c**) and downward (**d**). The points were fitted to a single exponential function plus a constant. The time constants are 5 ms (8.5 °C) at 20 mV (**c**) and 1.33 ms (13 °C) at −6 mV (**d**). (**a** from Chandler and Meves 1970a; **b, d** from Nonner 1980; **c** from Meves and Vogel 1977)

tial function up to 40 ms at 0 °C (the charge immobilizations shown in Fig. 7c, d are discussed in Sect. 3.2). On the other hand, two inactivation time constants are still discernible in frog (W. Schwarz 1979) and rat (J.R. Schwarz 1986) myelinated nerve at comparable low temperatures. The biphasic time course of sodium inactivation in frog myelinated nerve was studied in detail by Chiu (1977) and interpreted by second-order kinetics of the inactivation variable h with two separate inactivated, closed states following the open state. Alternatively, Ochs et al. (1981) suggest that the state of the nodal sodium channels follows the sequence: open-inactivated-open in order to explain the biphasic kinetics of inactivation during development and recovery.

A different explanation of the biphasic inactivation in frog myelinated nerve was given by Benoit et al. (1985) who suggest the presence of two types of sodium channels in the nodal membrane with fast and slow inactivation kinetics. This view is supported by the different effects of niflumic acid (Benoit et al. 1985), tetrodotoxin (Benoit and Dubois 1985), and guanidinium ions (Benoit and Dubois 1987) on the "fast" and "slow" sodium channels. Since the two types of sodium channels are assumed to be interconvertible (Benoit et al. 1985), they probably do not represent separate entities, and discriminating among homogeneous channel populations with complex inactivation characteristics seems to be rather difficult. Indeed, single sodium channels in neonatal cardiocytes with a high reopening probability during maintained depolarization produce fast *and* slow inactivation phases (Kohlhardt et al. 1988). A biphasic inactivation of sodium currents has also been observed in rat sodium channels type II expressed in *Xenopus* oocytes (Stühmer et al. 1987). This is probably the strongest argument in favour of the view that the nonexponential inactivation kinetics arise from uniform gating processes in all sodium channels rather than from different reactions in a population of heterogeneous channels.

3 Gating Currents

Gating currents from charge displacements in sodium channels have been measured in various axon preparations (see recent reviews by Bezanilla 1985; Meves 1989), in skeletal muscle (Collins et al. 1982b; Campbell 1983) and in *Xenopus* oocytes (Conti and Stühmer 1989). Since all voltage-dependent transitions contribute to the gating currents, measurement of these currents allows the detection and analysis of reactions in the gating process which take place before and after channel opening. This is an advantage over sodium currents which are restricted to the open channel states. The disadvantages of using gating currents are their tiny amplitudes, which complicate the measurement, and the difficulty of separating them from residual ionic currents. In addition, there was the suspicion that only part of the displacement current is related to gating in sodium channels and that an unknown fraction is due to other displaceable charges in the excitable membrane. These doubts expressed in earlier reviews (Ulbricht 1977; Almers 1978) have now been removed by recent measurements on *Xenopus* oocytes (Conti and Stühmer 1989) which show that the displacement currents measured in the presence of sodium channels are similar to gating currents in other preparations, while there is no charge displacement without channels. This indicates that the observed displacement currents arise almost exclusively from reactions in sodium channels. Hence, all electrophysiological procedures and chemical treatments known to eliminate certain fractions of the gating current, e.g. depolarizing prepulses or administration of local anaesthetics, block particular steps in the gating of sodium channels (Khodorov 1981; Keynes 1983). The chemical modification of gating currents in myelinated nerve and the resulting conclusions on the structure of sodium channels have been reviewed recently

(Meves 1989). In this review the main emphasis is on gating currents from unmodified sodium channels in order to reveal possible differences between channels in various membranes and to discuss the effects of inactivation on charge displacements in sodium channels.

3.1 Steady State Charge Distribution

Integration of the gating current during a long test pulse yields the steady state charge displacement which is denoted by Q_{on} for depolarizations (on-response) and by Q_{off} for repolarizations (off-response). The charges Q_{on} are a sigmoidal function of the membrane potential and reach a maximum value Q_{on}^{max} at large depolarizations (Fig. 8). This behaviour can be described by various models of the displacement of gating particles in the excitable membrane. The simplest possibility invokes first-order transitions of charges over a single energy barrier between two discrete states (Keynes and Rojas 1974). The steady state charge displacement Q at the membrane potential E of such a two-state model is described by the Boltzmann distribution:

$$Q(E) = \frac{Q^{max}}{\{1 + \exp[ze(E_0 - E)/kT]\}} \tag{1}$$

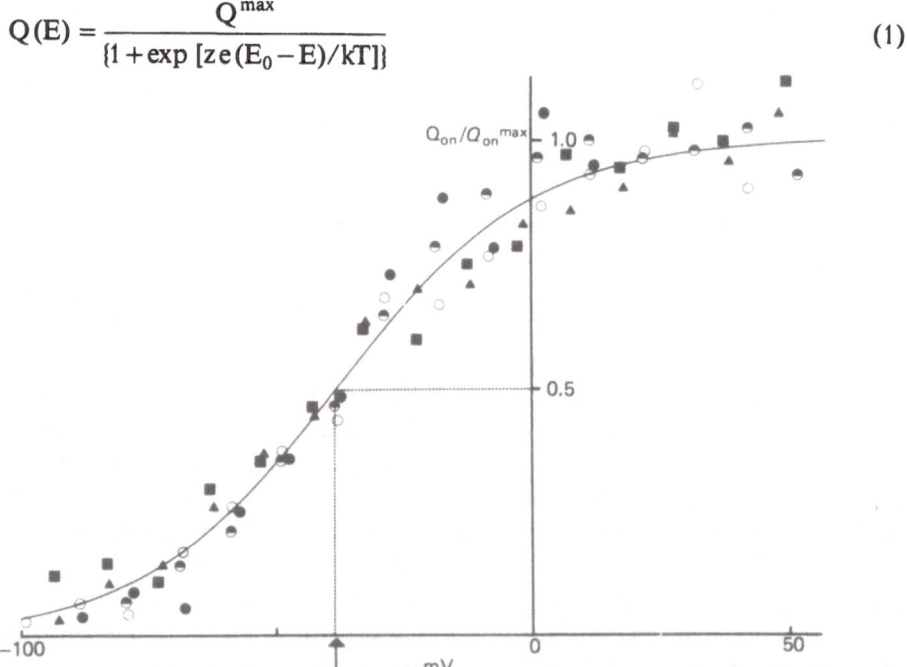

Fig. 8. Steady state charge distributions Q_{on} in frog skeletal muscle as a function of membrane potential. Different symbols denote results from different experiments. The Q_{on} values are normalized with respect to their maxima Q_{on}^{max} and fitted with the Boltzmann equation 1. Fit parameters: $z = 1.29$ and $E_0 = -38$ mV (*arrow*). Temperature 13°–19°C. (From Collins et al. 1982b)

E_0 is the midpoint potential at which $Q(E_0) = Q^{max}/2$, k is the Boltzmann constant, T the absolute temperature, e the elementary charge; z denotes the *effective* valence of the gating charges, i.e. their *actual* values multiplied by the fraction of the voltage drop across the membrane phase acting between the two discrete states. The effective valence can be obtained by fitting equation (1) to $Q(E)$ values or directly from the midpoint slope $zeQ^{max}/(4kT)$ of the $Q(E)$ curve. A related quantity is the number S of millivolts below the midpoint potential for an e-fold change of Q (called the limiting logarithmic potential sensitivity by Almers 1978), from which z is obtained through $z = kT/(eS)$. The effective valence of the gating charges is not only determined by the slope of the $Q(E)$ curve, but also depends on the model assumed for the charge displacement. For example, if instead of two-state transitions the particles may assume a larger number of possible states (Meves 1974; Neumcke et al. 1978), a higher valence is derived. Hence, fitting the steady state charge distribution to the Boltzmann equation (1) only gives a lower limit for the effective gating charge.

The Boltzmann distribution (1) is characterized by three parameters: Q^{max}, E_0, and z. Of these the maximum charge displacement Q^{max} is proportional to the number N of sodium channels contributing to the gating current. Since the proportionality factor is the product of the model-dependent valence z and the uncertain number of gating particles per sodium channel (see Sect. 2.3), estimates of N from measurements of Q^{max} are rather unreliable. The midpoint potential E_0 is equal to the membrane potential with zero electric field in the membrane phase and thus with equal distribution of gating

Table 1. Effective valencies z of gating charges in sodium channels

Tissue	z^a	Temperature (°C)	Reference
Squid giant axon	1.3	6 − 9	Keynes and Rojas (1974)
Squid giant axon	1.2	0 − 2	Meves (1974)
Myxicola giant axon	0.9	5 − 6	Rudy (1976)
Myxicola giant axon	1.6	4 − 5	Bullock and Schauf (1978)
Crayfish giant axon	1.8	5	Starkus et al. (1981)
Frog node of Ranvier	1.65	12	Nonner et al. (1975)
Frog node of Ranvier	1.7	10	Neumcke et al. (1976b)
Frog node of Ranvier	1.9	7.5 − 13	Dubois and Schneider (1982)
Xenopus node of Ranvier	1.3	6 − 25	P. Jonas (personal communication)
Frog skeletal muscle	1.3	19	Collins et al. (1982b)
Frog skeletal muscle	1.4	5	Campbell (1983)
Rat node of Ranvier	1.9	15 − 25	Chiu (1980)
Rat sodium channels type II in *Xenopus* oocyte	2.1	15	Conti and Stühmer (1989)

[a] Obtained by fitting the steady state charge distributions with the Boltzmann equation (1).

charges between the two states. In myelinated nerve E_0 is affected by the difficulty of determining the absolute value of the resting potential. Therefore, E_0 values from various axon preparations cannot be easily compared with each other.

This problem does not exist for the third parameter of the Boltzmann equation (1), the effective valence z of displaceable charges. Table 1 lists z values for sodium channels in various excitable membranes obtained from studies in different laboratories. There is some scatter in the reported z values between the extremes 0.9 and 2.1. The variations are probably not caused by the different temperatures, because no temperature dependence of the slope of the charge-distribution curve has been reported so far (Kimura and Meves 1979; Schauf and Bullock 1979; Collins and Rojas 1982; Jonas and Vogel 1988). Instead, there is the vague possibility that gating charges of rat sodium channels in peripheral nerve and of rat sodium channels expressed in *Xenopus* oocytes have a higher valence than those of other sodium channels. Thus, more accurate determinations of effective gating charges in different sodium channels may give further clues to channel diversity in the future.

3.2 Charge Immobilization

The gating charges of sodium channels displaced during depolarizations are restored to their original positions during sufficiently long repolarizations. However, after long pulses only part of the reverse charge displacement shows a rapid time course comparable to that of sodium deactivation, while the rest of the charge has a reduced mobility and flows back slowly over a period of several milliseconds. This inequality between the *rapid* charge displacements during (Q_{on}) and after (Q_{off}) longer depolarizing test pulses is called charge immobilization. The phenomenon has been described for giant axons of squid (Armstrong and Bezanilla 1977; Meves and Vogel 1977) and *Myxicola* (Bullock and Schauf 1979), for myelinated nerve (Nonner et al. 1978) and for skeletal muscle (Collins et al. 1982b; Campbell 1983). Figure 9 shows a comparison between the kinetics of sodium current and charge immobilization in the squid giant axon at the same test potential. The parallel time course of the inactivation phase of the sodium current and the Q_{off}/Q_{on} ratios suggest that charge immobilization and inactivation of sodium channels are closely related. This view is strengthened by the similar actions of toxins and chemical treatments on sodium inactivation and charge displacement: charge immobilization is prolonged when sodium inactivation is slowed, and charge immobilization is prevented when sodium inactivation is destructed (see reviews by Khodorov 1981; Keynes 1983; Meves 1989).

Despite these general agreements, there is no strict correlation between sodium inactivation and charge immobilization, e.g. in the squid giant axon the

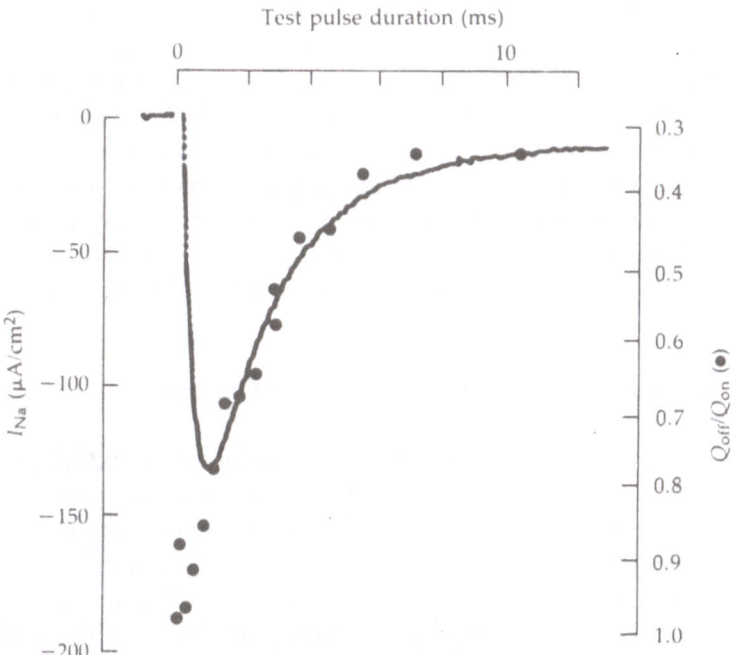

Fig. 9. Inactivation of sodium current I_{Na} (*solid line*) and immobilization of gating charge Q_{off}/Q_{on} (*filled circles*) in the squid giant axon at 0 mV, temperature 8 °C. (From Armstrong and Bezanilla 1977)

time constants of charge immobilization and sodium inactivation at 90 mV and 8 °C have been found to be 5 ms (Fig. 7 c) and 1.35 ms, respectively (Meves and Vogel 1977). The immobilization time constant was also reported to be larger than the inactivation time constant for *Myxicola* giant axons (Bullock and Schauf 1979) and for frog skeletal muscle (Collins et al. 1982 b). In muscle the charge immobilization has a time course similar to that of the slow inactivation process (see Sect. 2.4), while the classical h inactivation reaction is faster by one order of magnitude (Collins et al. 1982 a, b). In myelinated nerve the immobilization time constant is somewhat larger than that of the fast phase of sodium inactivation and much smaller than that of the slower inactivation process (Nonner 1980; see Fig. 7 b, d). Differences between the processes of sodium inactivation and charge immobilization are also observed in their stationary properties. Figure 9 shows inactivation of the sodium current to a low level whereas the Q_{off}/Q_{on} ratio reaches a larger limiting value of approximately 0.35. Hence, the steady state inactivation variable h_{∞} is smaller than the stationary value of Q_{off}/Q_{on} at a given test potential. A positive voltage shift of the normalized stationary off charge displacement with respect to the h_{∞} curve has also been found in *Myxicola* giant axons (Bullock and Schauf 1979) and in frog myelinated nerve (Nonner 1980).

The examples above do not invalidate the general finding that charge immobilization occurs and that it is intimately related to the process of sodium inactivation. Thus, the same gating charge is responsible for activation and inactivation (Bezanilla 1985). This is another indication for the existence of coupled activation and inactivation processes in sodium channels, since all gating schemes with independent gating reactions, e.g. the Hodgkin-Huxley formalism, predict that the charges displaced during activation should all return upon deactivation. Hence, charge immobilization narrows the number of possible reaction diagrams for the gating of sodium channels.

3.3 Slow Components of On-Gating Currents

As explained in Sect. 3.2, the main portion of the gating current reflects rapid charge displacements during the activation or deactivation of sodium channels. Additional slower gating current components arise during the on-response from the development of sodium inactivation and during the off-response from the recovery from inactivation. The amplitudes of these components are by one order of magnitude smaller than the total gating currents. For example, consider the contribution of an independent h inactivation reaction to the gating current. With an effective valence of four for this reaction (estimated from the slope of the h_∞ curve) and of six for the activation process (estimated from the slope of the m_∞^3 curve) the ratio of the initial h- and m-gating current amplitudes becomes $(4/6) \cdot (\tau_m/\tau_h)$. This expression holds for the on-response (transitions from $m_\infty = 0$, $h_\infty = 1$ to $m_\infty = 1$, $h_\infty = 0$) as well as for the off-response (transitions from $m_\infty = 1$, $h_\infty = 0$ to $m_\infty = 0$, $h_\infty = 1$). In both cases, the h- versus m-gating current amplitudes depend on the activation and inactivation time constants τ_m, τ_h at the respective test potentials. Examples of τ_m/τ_h values at membrane potentials near 0 mV are 0.21 for the squid giant axon at 6.3 °C (Hodgkin and Huxley 1952), 0.12 for the crayfish giant axon at 8 °C (Swenson 1983), 0.10 for frog and 0.12 for rat myelinated nerve at 20 °C (Neumcke et al. 1987). If the processes of sodium activation and inactivation are coupled, the effective valence of inactivation is less than four (see Sect. 5) and the relative h-gating current amplitude becomes even smaller. A further complication is that the time constant τ_h of sodium inactivation is of the same order as the time constant τ_n of potassium activation (Hodgkin and Huxley 1952). Hence, it is difficult to discriminate between slow h-gating current components from possible charge displacements and residual ionic currents of potassium channels.

There are two possible methods of showing that slow gating current components might be related to inactivation reactions in sodium channels. The first method is to establish an agreement between the time constants of the components and the inactivation time constant τ_h over a wide range of mem-

Fig. 10a–c. Slow components of on-gating currents in the crayfish giant axon (a) and in frog myelinated nerve (*arrow* in b). The slow component in myelinated nerve is abolished by 7 µM Anemonia toxin ATX II in the external solution (c). Test pulses to 20 mV (a) and 30 mV (b, c); temperatures 8 °C (a) and 15 °C (b, c). The same units of abscissa and ordinate are used in b, c and a resting potential of −70 mV is assumed. (a from Swenson Jr. 1983; b, c from Neumcke et al. 1985)

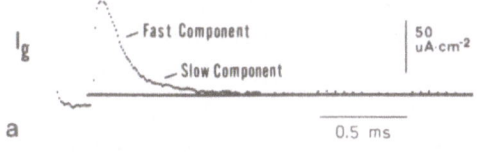

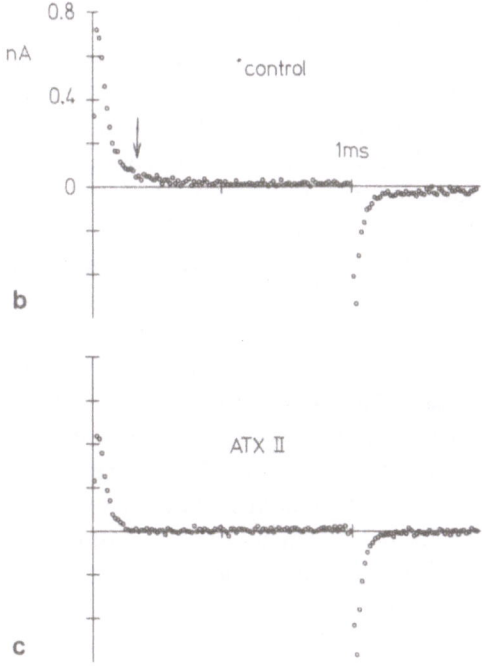

brane potentials. The second method is to show that *specific* modifiers of sodium inactivation also alter the slow components in the gating currents. The requirements seem to have been fulfilled in some recently published gating current studies. An example is illustrated in Fig. 10a which shows on-gating currents in crayfish giant axon with a clear separation between fast and slow components (Swenson Jr. 1983). The time constant of the slow phase was found to be similar to τ_h, between 0 and 40 mV, and, therefore this phase was attributed to the h inactivation process. Slow components of the on-charge movement have also been described for frog myelinated nerve (Dubois and Schneider 1982). However, the reported time constants of the components are smaller than τ_h. Nevertheless, at least some of the slow on-gating current components in myelinated nerve seem to be related to the h-gating process because they can be reduced by various substances which slow sodium inactivation. This has been reported for the venom of the scorpion *Buthus eupeus* and for the alkaloid aconitine by Krutetskaya et al. (1978), for the scorpion venom of *Leiurus quinquestriatus* by Nonner (1979) and for the toxin ATX II of *Anemonia sulcata* by Neumcke et al. (1985). The effect of ATX II is illustrated in Fig. 10b, c. Of these agents, scorpion venoms (Koppenhöfer and Schmidt 1968) and possibly also aconitine (Schmidt and Schmitt 1974) not only slow the process of sodium inactivation but also delay the activation of potassium channels as well. Hence, the abolishment of slow on-gating current components observed after application of these substances could be due to modifications of gating processes in sodium or potassium channels. The toxin ATX II, on the other hand, prolongs sodium inactivation but does not affect the kinetics of potassium currents in myelinated nerve (Bergman et al. 1976). The absence of slow gating current components with ATX II (Fig. 10c) is, therefore, good evidence in support of the view that the components observed under control conditions (arrow in Fig. 10b) are indeed related to the inactivation of sodium channels.

Do h-gating current components allow discrimination between different reaction schemes for the sodium channel? Unfortunately, all gating diagrams predict slow components in the gating currents, even those in which inactivation is strictly coupled to activation and inactivation does not have its own voltage dependence (see Sect. 5.2). The model with coupled activation and inactivation also accounts for the diminution of the slow components by inactivation modifiers (Fig. 20). However, a model-dependent property of the inactivation component is its relative initial amplitude: the ratio between the initial h- and m-gating current amplitudes in the on-response is of the order of 0.1 for independent gating reactions (assuming $\tau_m/\tau_h = 0.15$, see Sect. 3.3), but is smaller with coupled activation and inactivation (amplitude ratio 0.02 as obtained from the calculated gating currents in Fig. 20). These theoretical values could help to identify and interpret slow inactivation components in gating current measurements.

4 Single Channel Measurements

4.1 Sodium Channel Conductance

Early estimates of the conductance γ of single sodium channels were obtained from the analysis of sodium current fluctuations (for a review see Neumcke 1982). Higher γ values were obtained for myelinated nerve fibres than for squid giant axon: Determinations of γ values for the squid giant axon in different studies were 4 pS at 9 °C (Conti et al. 1975), 3.5 pS at 3°–6 °C (Llano and Bezanilla 1984), and 4.4 pS at 3.5°–5 °C (Bekkers et al. 1986), whereas myelinated nerve γ values were found to equal 7.9 pS at 13 °C (Conti et al. 1976) and 6.4 pS at 2°–5 °C (Sigworth 1980) in frog and 14.5 pS at 20 °C

(Neumcke and Stämpfli 1982) in rat. Since the Na$^+$ concentration of sea water is four to five times higher than that of Ringer's solution, the difference between the channel conductances in the giant axon and in myelinated nerve becomes even more pronounced.

The first measurements of currents through single sodium channels in rat myoballs (Sigworth and Neher 1980), rat myotubes (Horn et al. 1981 b), bovine chromaffin cells (Fenwick et al. 1982), mouse neuroblastoma cells (Quandt and Narahashi 1982; Nagy et al. 1983), and rat heart myocytes (Cachelin et al. 1983) have established a range of 10–20 pS for the conductance of vertebrate sodium channels. In addition, the results of these experiments could give an explanation for the lower conductance in the squid giant axon. Thus, Yamamoto et al. (1984) observed a rapid and voltage-dependent block of single sodium channels in neuroblastoma cells caused by extracellular Ca^{2+} ions, resulting in a decrease in the apparent channel conductance at increased Ca^{2+} concentrations. A similar voltage-dependent block of open sodium channels caused by extracellular Ca^{2+} ions has been observed in single canine cardiac Purkinje cells (Sheets et al. 1987). These findings suggest that the lower channel conductance in the squid giant axon is mainly caused by the higher extracellular Ca^{2+} concentration in sea water compared to Ringer's solution. This view is supported by a direct measurement of 14 pS for the conductance of squid sodium channels in solutions containing no divalent cations (Bezanilla 1987). Thus, it appears that the different conductance values in invertebrate and vertebrate sodium channels under physiological conditions can be attributed to differences in the ionic compositions of the extra- and intracellular solutions, and therefore there is no necessity to invoke species-dependent pores in the sodium channel protein.

Sodium channels expressed in *Xenopus* oocytes have a conductance comparable to that of natural channels in adult and cultured cells (Stühmer et al. 1987; Sigel 1987 a, b). Such an agreement cannot as yet be established for purified sodium channels reincorporated into artificial lipid membranes because the channels are normally modified by the alkaloids batrachotoxin (BTX) or veratridine to prolong channel opening. Two exceptions are found in the early investigations by Hanke et al. (1984) and Rosenberg et al. (1984) which were performed on unmodified sodium channels reconstituted into planar lipid bilayers or into multilamellar lipid vesicles, respectively. Obviously, the properties of the incorporated unmodified sodium channels are different from those in their natural environments (see also Duch and Levinson 1987). Thus, the 25- and 150-pS channels described by Hanke et al. (1984) exhibit gating kinetics which are much slower than those of sodium channels in biological membranes. Treatment with BTX created channels of larger conductance in the study by Rosenberg et al. (1984), whereas BTX-modified sodium channels in adult and cultured cells have a reduced conductance (Khodorov et al. 1981; Quandt and Narahashi 1982). It is also difficult to compare the conductances of modified sodium channels in biological membranes with those in lipid bilayers because the measurements were carried out using solutions of different ionic compositions. For example, BTX-modified sodium channels in neuroblastoma cells have a conductance of only 2 pS when the extracellular solutions contain 125 mM Na and 1.8 mM Ca (Quandt and Narahashi 1982) and 10 pS with $[Na]_0 = 130$ mM, $[Ca]_0 = 0.5$ mM (Huang et al. 1984). For reconstituted BTX-modified sodium channels the following conductances were determined using symmetrical solutions on both sides of the lipid membrane: 30 pS in 500 mM $NaCl + 0.15$ mM $CaCl_2 + 0.1$ mM $MgCl_2$ (Krueger et al. 1983), 25 pS in 500 mM NaCl (Green et al. 1984), 20 pS in 200 mM NaCl (Moczydlowski et al. 1984), 25 pS in 500 mM NaCl and 18 pS in 100 mM NaCl (Recio-Pinto et al. 1987), and 26 pS in 500 mM NaCl (Duch et al. 1988). The higher conductance of BTX-modified sodium channels in artificial lipid membranes could simply be due to the solutions in the bilayer experiments which were symmetrical and of reduced or vanishing concentrations of divalent cations or it could be a genuine difference reflecting the influence of the foreign lipid environment on the sodium channel protein. The same ambiguity arises when the conductances of veratridine-modified sodium channels in biological membranes and lipid bilayers are compared with each other. Again, the channel conductances of 5–6 pS in *Xenopus* oocytes (Sigel 1987 b), 4 pS in neuroblastoma cells (Barnes and Hille 1988), and 13 pS (Recio-Pinto et al. 1987) and 10 pS (Garber and Miller 1987) in lipid membranes were measured in different solutions, and therefore it cannot be determined whether modified sodium channels have the same or different conductances in biological and artificial membranes.

4.2 Low-Conductance Channels and Subconductance States

Lowered conductances in sodium channels are not only observed with increased Ca concentrations in the solutions or after treatment with alkaloids, but they are also found in physiological solutions and under unmodified conditions. These low-conductance sodium channels frequently occur in preparations from cardiac muscle together with sodium channels of normal conductance (Kunze et al. 1985; Scanley and Fozzard 1987; Kohlhardt et al. 1987). As an example, Fig. 11 a (top) shows currents in a normal and a low-conductance sodium channel of a myocyte prepared from rat ventricle cells. These two sodium channels produce different peaks in the amplitude histogram (Fig. 11 a, bottom), but there are no large differences in their gating kinetics. Similarly, Scanley and Fozzard (1987) found the same kinetics for normal and low-conductance sodium channels in canine cardiac Purkinje cells. On the other hand, Kohlhardt et al. (1987) reported that low-conductance sodium channels in rat cardiocytes exhibit slower inactivation, a longer and voltage-

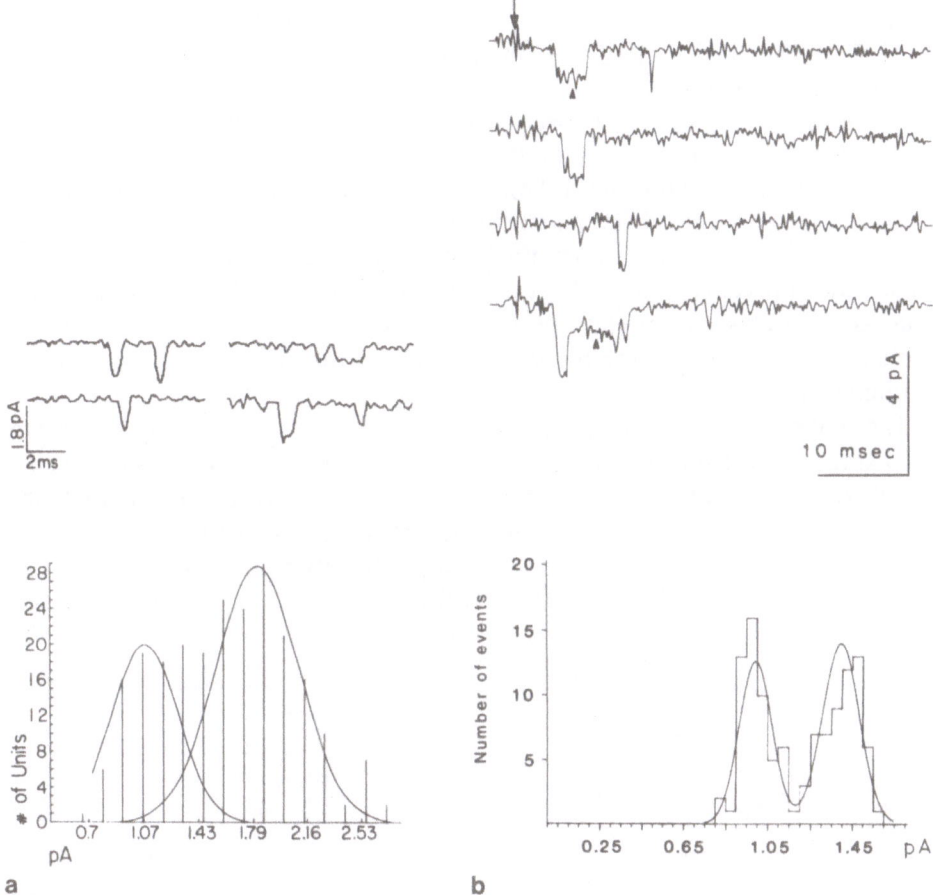

Fig. 11a,b. Low-conductance sodium channels in a cardiac myocyte (**a**) and in a muscle myoblast (**b**). The *upper* parts show single channel records which illustrate two populations of sodium channels. Currents from the low-conductance channel in **b** are indicated by *upward arrowheads*. The *lower* parts give the corresponding amplitude histograms and the fits with double Gaussian curves. Test voltages 0 mV (**a**) and −40 mV (**b**). (**a** from Kunze et al. 1985; **b** from Weiss and Horn 1986)

dependent open time, and sometimes even an altered ionic selectivity with respect to normal sodium channels.

In rat skeletal muscle the sensitivity of sodium currents to tetrodotoxin (TTX) increases during development (Harris and Marshall 1973; Frelin et al. 1984). This has been explained by a decrease in the number of TTX-resistant and a concomitant creation of TTX-sensitive sodium channels. These two postulated types of sodium channels were identified in single channel experiments on muscle cultures of neonatal rats (Weiss and Horn 1986). Figure 11b shows currents in normal and low-conductance sodium channels of a myoblast cell and the corresponding amplitude histogram. Since the low-conductance channels are resistant to TTX, they generate muscle excitability in the first stages of development. Probably, the same low-conductance, TTX-resistant sodium channel coexists with the normal TTX-sensitive channel in denervated rat muscle (Harris and Thesleff 1971; Pappone 1980) and in cultured mammalian muscle myoballs (Ruppersberg et al. 1987; Pröbstle et al. 1988; Ruppersberg and Rüdel 1988). It has been inferred from experiments on *Xenopus* oocytes that different messenger ribonucleic acids (mRNAs) code for the two types of sodium channels in innervated and denervated muscles of cat and rat (Parker et al. 1988), while functional TTX-sensitive rat brain sodium channels can already be expressed by the injection of the mRNA which codes exclusively for the large α channel peptide (Noda et al. 1986b; Stühmer et al. 1987; Suzuki et al. 1988).

Low-conductance sodium channels which represent separate entities have to be distinguished from subconductance states in one sodium channel. It is difficult to discriminate between different channels and different conductance levels in a single channel, but frequent direct transitions between higher current levels and the zero-current baseline make the latter possibility more likely. On the other hand, different TTX sensitivities of the various conductance

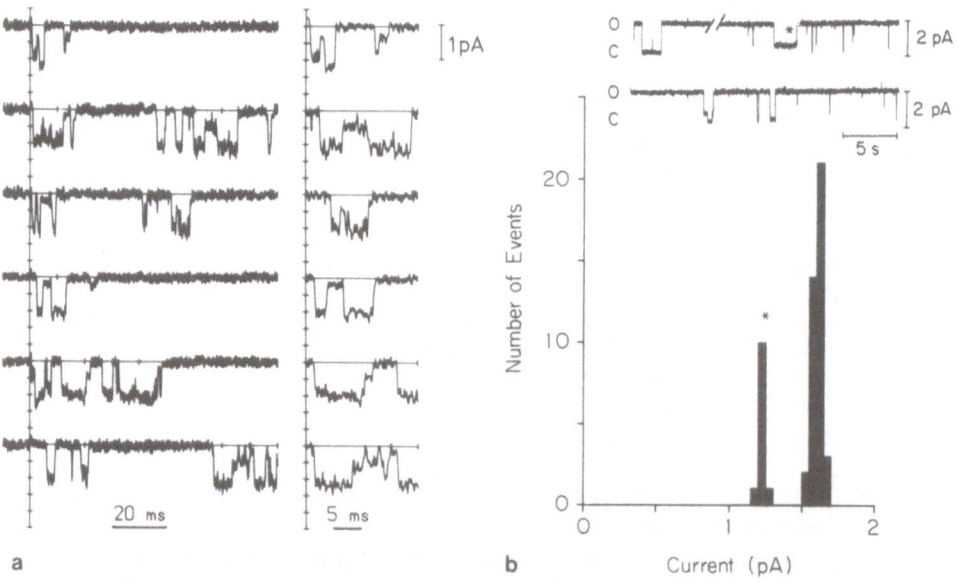

Fig. 12a, b. Subconductance states of sodium channels in a neuroblastoma cell (**a**) and in a planar lipid bilayer membrane (**b**). Sodium channels were modified by treatment with chloramine-T (**a**) or with batrachotoxin (**b**). **a** Single channel currents (plotted downward) during 90-ms depolarizations of 80 mV (*left*). Segments with sublevels are shown on an expanded time scale (*right*). Ordinates are in units of 0.5 pA. **b** Single channel currents (plotted upward from the closed c to the open o state) in 500-mM NaCl solutions at a membrane potential of 63 mV (*top*), and the corresponding amplitude histogram (*bottom*). The subconductance state is marked by asterisks. (**a** from Nagy 1987b; **b** from Green et al. 1987)

states suggest the presence of separate sodium channels (Nilius et al. 1989). While some indications have been presented for the existence of subconductance states in unmodified sodium channels of neuroblastoma cells (Nagy et al. 1983), the most convincing evidence for multilevel sodium channels comes from recent studies on modified sodium channels. In particular, subconductance states have been detected in sodium channels of neuroblastoma cells treated with the insecticide deltamethrin (Chinn and Narahashi 1986), with chloramine-T, sea anemone toxin ATX II or scorpion toxin (Nagy 1987b, 1988; Meves and Nagy 1989), in sodium channels in skeletal and heart muscle modified by the cardiotonic compound DPI 201-106 (Patlak 1988; Nilius et al. 1989) and in sodium channels which have been incorporated into lipid bilayer membranes and modified by BTX (Green et al. 1987; Recio-Pinto et al. 1987; Duch et al. 1988). Examples are shown in Fig. 12. Interestingly, the sublevel conductances of modified sodium channels in neuroblastoma cells are not randomly distributed but are close to one-quarter and one-half of the main conductance state. In addition, a one and one-quarter conductance level has been observed (Nagy 1987b). It remains to be shown whether these one-quarter conductance steps are generated by a conformational change in one of the four homologous units in the sodium channel protein (Noda et al. 1984).

4.3 Statistical Properties of Single Sodium Channels

The fast activation and slow inactivation of macroscopic sodium currents during depolarizing test pulses can be described with numerous reaction schemes for the gating of sodium channels (see Sect. 5). The analysis of the statistical opening and closing of single sodium channels offers the possibility of selecting among various reaction diagrams because different gating schemes predict different single channel behaviours (French and Horn 1983; Aldrich 1986). There are two reasons, however, why a definite state diagram for sodium channels can not as yet be derived from the statistical properties of single sodium channels. Firstly, closely related schemes show only minor differences in the predicted gating behaviour and discrimination among models requires extensive computations using "maximum likelihood procedures". This is only feasible for reaction diagrams having up to five states (Horn and Vandenberg 1984). Secondly, the kinetics of a single sodium channel may not be determined by one reaction diagram with time-independent rate constants, but there may be slowly interchangeable gating modes, as observed in sodium channels of rat heart and frog skeletal muscle (Patlak and Ortiz 1985, 1986; Patlak et al. 1986). In the following analysis the switching of modes between multiple kinetic diagrams is neglected and the statistical properties of sodium channels are described by a single reaction scheme. In particular, measured

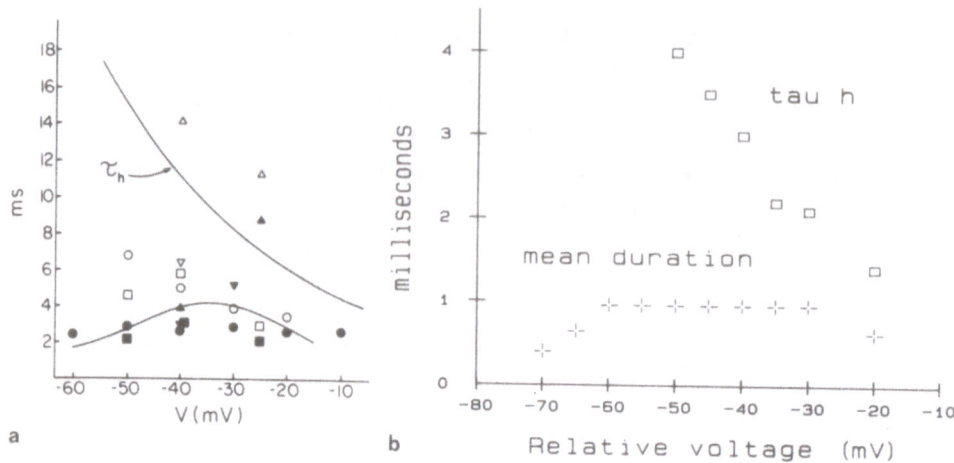

Fig. 13a, b. Time constants of sodium inactivation (τ_h) and mean durations of open sodium channels in outside-out patches of GH_3 (a) and in cell-attached patches of neuroblastoma cells (b). a The τ_h curve represents an exponential fit to inactivation time constants obtained from macroscopic sodium currents and from averaged single channel records. The lower curve represents predicted open times calculated from the average linear regression values of β_A and β_1 (see Fig. 14a, b). *Filled symbols* denote measured open times and *open symbols* denote burst durations; V is the membrane potential, temperature 9.3 °C. b The τ_h values were obtained by an exponential fit to the decline of averaged single channel records. Voltages are relative to the (unknown) resting potential of the cell, temperature 11 °C. (a from Vandenberg and Horn 1984; b from Aldrich and Stevens 1987)

values of the duration of the open channel state and of the inactivation and deactivation rate constants are discussed and compared with the predictions of gating models with independent or coupled activation and inactivation.

According to the m^3h description of sodium currents, the inactivation rate constant β_h predominates over the rate constants α_h and β_m at large depolarizations (Hodgkin and Huxley 1952). Consequently, the time constant of sodium inactivation τ_h which equals $1/(\alpha_h + \beta_h)$, and the mean channel open time τ_0, which equals $1/(3\beta_m + \beta_h)$, approach the same value $1/\beta_h$ at increasing membrane potentials (see Fig. 21 c). In the first studies of single sodium channels in rat myoballs (Sigworth and Neher 1980) and in tunicate eggs (Fukushima 1981) the relationship between τ_h and τ_0 was found to be at least qualitatively fulfilled. However, in subsequent investigations the open times were always significantly smaller than the inactivation time constants and τ_0 did not reach a constant value at increased membrane potentials as expected from the m^3h formalism (Aldrich et al. 1983; Nagy et al. 1983; Vandenberg and Horn 1984; Aldrich and Stevens 1987). Figure 13 shows τ_0 and τ_h values for excised "outside-out" patches of GH_3 cells (Fig. 13a) and for cell-attached patches of neuroblastoma cells (Fig. 13b). A comparison of the experimental results with model predictions (Fig. 21 c, d) reveals that a reaction scheme with coupled activation and inactivation can account for the difference between τ_0 and τ_h over a large voltage range.

The inactivation time constants and open times in Fig. 13a, b exhibit similar voltage dependencies, but their absolute values are very different. For example, the plateau of the open time curve is near 4 ms in outside-out patches of GH_3 cells and approximately 1 ms in cell-attached patches of neuroblastoma cells. Since the experiments were performed at almost the same temperature, the difference must be related either to the cell type or to the patch configuration. The latter possibility is suggested by open times of the order of 1 ms for cell-attached patches from neuroblastoma cells (Fig. 13b) and from rat ventricular cells (Kunze et al. 1985), while larger open times have been reported for excised patches of a variety of cells, i.e. GH_3 cells (Fig. 13a), neuroblastoma cells (Nagy et al. 1983; Nagy 1987a) and dorsal root ganglion neurons (Carbone and Lux 1986). Direct evidence of the alteration of sodium channels by excision from intact cells comes from a comparative study on GH_3 cells in the cell-attached and inside-out configurations (Horn and Vandenberg 1986). After excision the open time became longer, the probability of channel opening during depolarization increased and the openings occurred no longer at random but in bursts. A similar modification of channel gating by membrane excision has been reported for cardiac sodium channels (Nilius 1988). Horn and Vandenberg (1986) attributed the changes in gating behaviour to the replacement of the cytoplasm by the internal CsF solution. Moreover, cytoplasmatic microtubules linked to sodium channels influence the sodium channel properties in intact cells (Sakai et al. 1985) and could be destroyed by the excision procedure.

The closure of open sodium channels can occur either by transition to the inactivated state or by deactivation to the resting closed state. Of particular interest are the voltage dependencies of the two closing reactions because they allow determinations of the charge displacements in the individual gating processes. The first step in the determination of the inactivation (β_I) and deactivation (β_A) rate constants is to measure the channel open time τ_0, which is equal to the reciprocal of the sum of both rates: $\tau_0 = 1/(\beta_I + \beta_A)$. To obtain β_I and β_A separately the following three procedures have been suggested:

1. As outlined by Stevens (1986) and Aldrich and Stevens (1987), the probability F of closing by inactivation equals $\beta_I/(\beta_I + \beta_A)$ and it depends on the experimentally obtainable probabilities that a channel reopens during depolarization and that it performs a direct transition from the resting closed states to the inactivated state. Thus β_I and β_A can be calculated from F and τ_0 through $\beta_I = F/\tau_0$ and $\beta_A = 1/\tau_0 - \beta_I$. A disadvantage of this method is that it only allows the determination of a lower limit for F at small depolarizations and that the actual value of β_I may be larger and that of β_A smaller than the calculated rate constants in this voltage range.

Moreover, the estimated values of F are restricted to state diagrams in which all direct transitions between the resting closed states and the inactivated state have the same probability.

2. With maximum likelihood methods the rate constants of a given reaction scheme can be adjusted to give the best agreement with experimental single channel records (Horn and Vandenberg 1984). The resulting rate constants critically depend on the selected gating model, but the advantages over method 1 are that absolute values for the inactivation and deactivation rates can be obtained at all voltages and that no restrictions on individual rate constants have to be imposed.

3. A pharmacological approach to determine β_I and β_A is to use substances which selectively eliminate the inactivation reaction ($\beta_I = 0$). The open time $\tau_0 = 1/(\beta_I + \beta_A)$ is then expected to increase and the difference between the reciprocal open times $1/\tau_0$ before and after drug application should be equal to the rate β_I under normal conditions. Indeed, a prolongation of the open time of sodium channels — and in most of the preparations an increased burst activity — has been observed after treatment with the inactivation modifiers N-bromoacetamide (Patlak and Horn 1982), toxin ATX II of *Anemonia sulcata* (Schreibmayer et al. 1987), papain (Quandt 1987) and BTX (Quandt and Narahashi 1982; Grant and Starmer 1987). For BTX-modified sodium channels in cardiac myocytes β_A was found to decline at increasing depolarizations, while in normal channels the sum $\beta_I + \beta_A$ assumed a minimum. Grant and Starmer (1987) have concluded from these results that β_I for normal sodium channels increases at more positive membrane potentials and that this reflects a voltage dependence of the inactivation reaction. However, BTX not only eliminates inactivation but also has profound effects on sodium activation (for an extensive review on BTX see Khodorov 1985). Hence, experiments with BTX are not appropriate for reliable determinations of β_I and β_A.

Values of β_I and β_A at different voltages are given in Fig. 14. The data shown in Fig. 14a, b are obtained from outside-out patches of GH$_3$ cells using the maximum likelihood method applied to a "basic model" in which inactivation can occur either from the open or from the last of three closed states (Vandenberg and Horn 1984). The results reveal similar voltage dependencies for β_A and β_I (equivalent gating charges 1.30 for deactivation and 1.92 for inactivation). On the other hand, in Fig. 14c almost the entire voltage dependence is attributed to the deactivation closing reaction (equivalent gating charge 3.42), while inactivation is only affected to a small extent by voltage (equivalent gating charge 0.46). These results were obtained from cell-attached patches of neuroblastoma cells employing procedure 1 to determine β_I and β_A (Aldrich and Stevens 1987). Probably, the different voltage dependencies of deactivation and inactivation shown in Figs. 14a–c are not due to

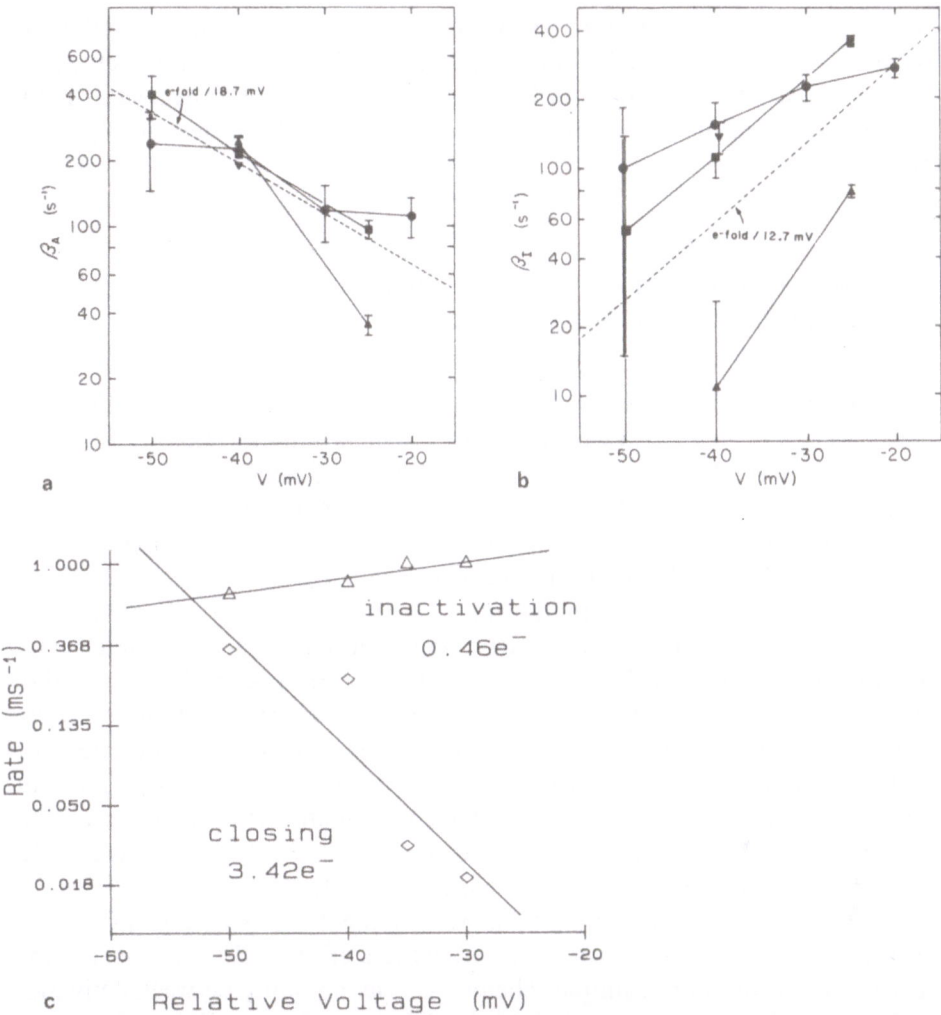

Fig. 14a–c. Voltage dependence of deactivation (β_A, closing) and inactivation (β_I) rate constants in outside-out patches of GH_3 (**a, b**) and in cell-attached patches of neuroblastoma cells (**c**). Note logarithmic scaling of ordinates. **a, b** The *dashed lines* have voltage dependencies of e-fold/18.7 mV (equivalent gating charge 1.30) for deactivation and e-fold/12.7 mV (equivalent gating charge 1.92) for inactivation. V is the membrane potential, temperature 9.3 °C. **c** Voltages are relative to the (unknown) resting potential of the cell, temperature 11 °C. (**a, b** from Vandenberg and Horn 1984; **c** from Aldrich and Stevens 1987)

the different methods used to analyse single channel records but are related to the different patch configurations. Thus, it appears that excision of membrane patches from intact cells not only affects the probability and the duration of the open channel state (see above), but also influences the deactivation and inactivation gating reactions. Comparison of Fig. 14 with the model calculations in Fig. 21a, b suggests that reaction diagrams with little inherent voltage dependence for the inactivation step − strict or partial coupling between activation and inactivation − are appropriate for sodium channels in intact cells and that channels in isolated membrane patches behave as if the processes of sodium activation and inactivation had been decoupled by the excision procedure.

5 A Reaction Diagram for the Sodium Channel
With Coupled Activation and Inactivation

In the previous sections several properties of sodium channels were discussed which cannot be described with the conventional Hodgkin-Huxley equations. Examples are the large m_∞ shifts in neuroblastoma cells produced by inactivation modifiers (Sect. 2.2), the delayed development of sodium inactivation and the differences between the τ_h and τ_c inactivation time constants in some giant axons (Sect. 2.4), the immobilization of gating charges (Sect. 3.2), and the statistical properties of single sodium channels in intact cells (Sect. 4.3). All cases indicate that the independence between sodium activation and inactivation as presumed in the Hodgkin-Huxley equations is no longer fulfilled under the conditions of the described experiments. Therefore, numerous reaction schemes have been suggested in which the gating processes of activation and inactivation are coupled to each other to various degrees (examples are given in the papers of Hoyt 1963, 1968; Goldman 1975; Jakobsson 1976; Bezanilla and Armstrong 1977; Armstrong and Bezanilla 1977; Armstrong 1978; Conti et al. 1980; Nonner 1980; Bean 1981; Khodorov 1981; Keynes 1983; Aldrich and Stevens 1984; Horn and Vandenberg 1984; Fishman 1985). Instead of examining the merits and deficiencies of all of these proposals, this review concentrates on a simple scheme − strict coupling between activation and inactivation with voltage-independent inactivation rates − which represents the opposite extreme to that represented by the Hodgkin-Huxley formalism, i.e. independence between activation and inactivation with voltage-dependent inactivation rates. The reaction diagram for the gating scheme discussed in this section is:

$$C_1 \underset{b}{\overset{3a}{\rightleftharpoons}} C_2 \underset{2b}{\overset{2a}{\rightleftharpoons}} C_3 \underset{3b}{\overset{a}{\rightleftharpoons}} O \underset{d}{\overset{c}{\rightleftharpoons}} I \ . \tag{2}$$

Here, C_1, C_2 and C_3 denote three resting closed states in the sodium channel. The inactivated closed state I can only be reached through the open state O which implies strict coupling between sodium activation and inactivation. The sequence 3-2-1 of the forward and reverse activation rate constants a and b was chosen in accordance with the conventional m^3 formulation of sodium activation. In contrast to the Hodgkin-Huxley inactivation parameters, it is assumed that the rate constants c and d in scheme 2 are voltage-independent, which implies that sodium inactivation derives its voltage dependence entirely from the preceding activation (Bezanilla and Armstrong 1977). Special features of this and similar sequential schemes have been described earlier (Bezanilla and Armstrong 1977; Armstrong 1978; Keynes et al. 1982; Neumcke et al. 1985), and it has been pointed out that such models cannot account for the statistical properties of single sodium channels in rat myotubes (Horn et al. 1981 a) and in GH_3 cells (Horn and Vandenberg 1984). Nevertheless, gating scheme 2 is used to illustrate the effects of coupling between sodium activation and inactivation and the consequences of voltage-independent inactivation rates on the properties of macroscopic sodium currents, gating currents, and single sodium channels. The values of the rate constants a, b, c and d and the voltage dependence of a and b are chosen to give an approximate description of sodium currents in frog myelinated nerve at 15 °C (Neumcke et al. 1985) and in rat nerve at 20 °C (Neumcke et al. 1987).

5.1 Description of Macroscopic Sodium Currents

The amplitude of the sodium current is proportional to the probability p_0 of the open state O. This parameter was calculated numerically by solving the fourth-order differential equations following from reaction scheme 2. With the rate constant values $a = 12$ ms^{-1}, $b = 1.8$ ms^{-1}, $c = 2.3$ ms^{-1} and $d = 0.01$ ms^{-1}, a good description of the kinetics of sodium currents in a frog node at $E = -10$ mV can be achieved (control curves in Fig. 15). The slowed and incomplete sodium inactivation after treatment with *Anemonia* toxin II is simulated with the same values for a and b and with $c = 1.1$ ms^{-1} and $d = 0.2$ ms^{-1} (interrupted curves in Fig. 15).

Since the inactivation rate constants c and d in scheme 2 are assumed to be voltage-independent, the values $c = 2.3$ ms^{-1} and $d = 0.01$ ms^{-1} are used for normal sodium channels at all potentials, while the activation rate constants a and b are varied to account for the voltage dependence of the sodium currents. The ratio a/b is specified by the h_∞ curve. In terms of scheme 2 the stationary probability $1 - h_\infty$ of the inactivated state I is given by the equation:

$$1 - h_\infty = \frac{c/d}{(1 + b/a)^3 + c/d} \tag{3}$$

and an approximate description of the experimental h_∞ curve of myelinated nerve is achieved with:

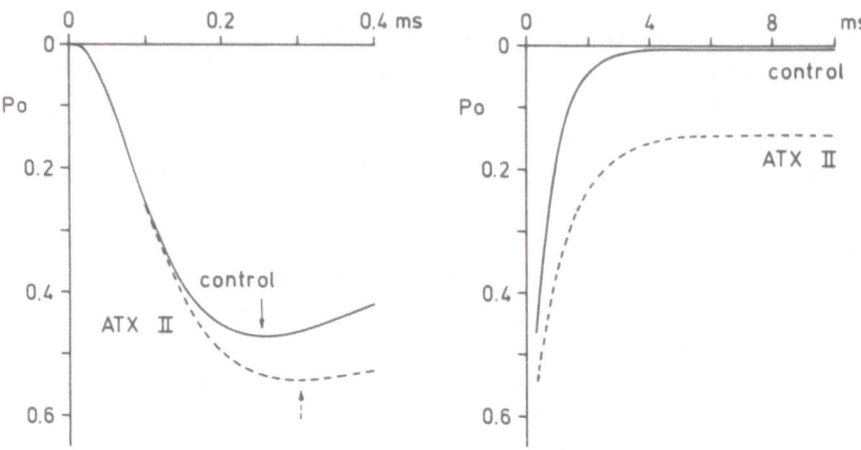

Fig. 15. Simulation of sodium currents in frog nerve at $E = -10$ mV under control conditions (*full curves*) and after treatment with *Anemonia* toxin ATX II (*interrupted curves*). The *left-hand* graph shows the activation phase on an extended time scale, the *right-hand* graph shows the inactivation phase after the peak of the sodium current. The probabilities p_0 of the open state O (plotted downwards) were calculated from scheme 2 with the rate constants (in ms^{-1}): $a = 12$, $b = 1.8$, $c = 2.3$, $d = 0.01$ (control) and $a = 12$, $b = 1.8$, $c = 1.1$, $d = 0.2$ (ATX II). Probability of state $C_1 = 1$ as initial condition. Times to the peak of the sodium current are marked by *arrows*

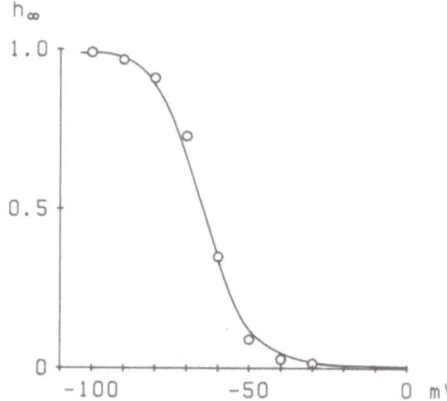

Fig. 16. Comparison of calculated and measured h_∞ values. The curve was computed from equation 3 with $c/d = 230$ and b/a from equation 4. The symbols are h_∞ values of rat myelinated nerve fibres with an assumed resting potential of -70 mV (from Neumcke et al. 1987)

$$\frac{b}{a} = \exp\left(\frac{E + 35 \text{ mV}}{-18 \text{ mV}}\right) \tag{4}$$

(see Fig. 16). If a and b are interpreted as the forward and backward rate constants of a particle hopping over an energy barrier, an effective valency of 1.4 elementary charges follows from the relationship in equation 4. Thus the slope and midpoint potential of the experimental h_∞ curve can be predicted with a reasonable value of the gating charge and with voltage-independent in-

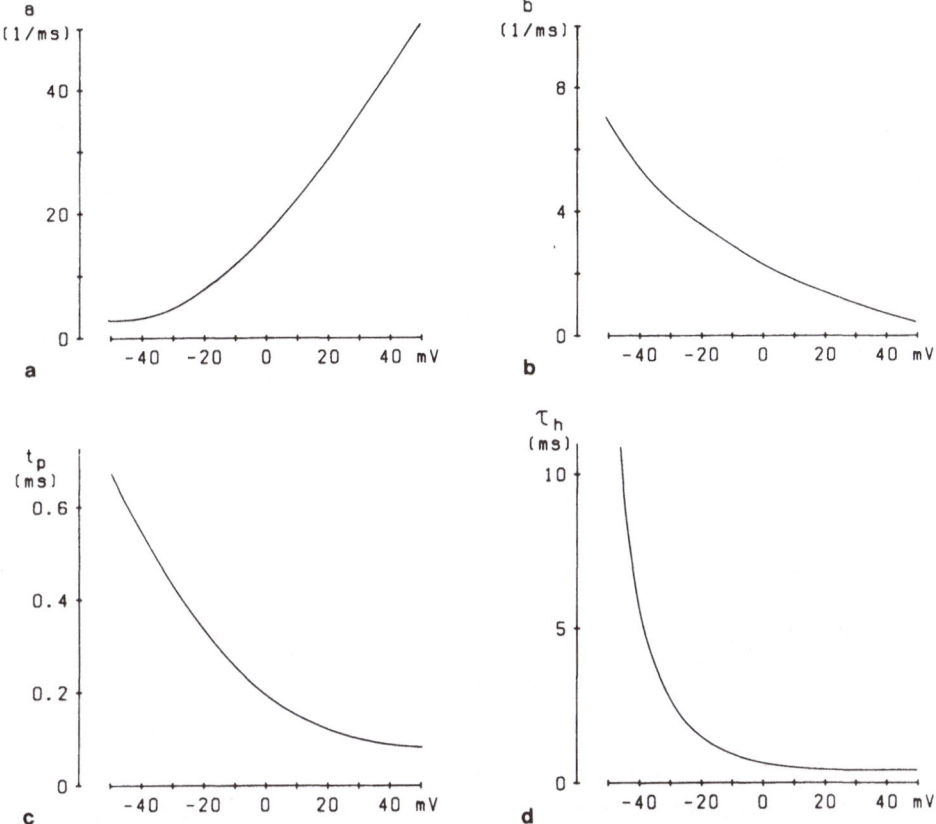

Fig. 17a–d. Voltage dependence of activation rate constants a (**a**) and b (**b**), of times t_p to peak sodium current (**c**), and of inactivation time constants τ_h (**d**). Values t_p and τ_h were calculated with parameters a, b (from **a, b**) and with voltage-independent inactivation rates $c = 2.3\ ms^{-1}$ and $d = 0.01\ ms^{-1}$. The rates and times refer to sodium currents in rat myelinated nerve at 20 °C

activation rates. The final step in the specification of the rate constants is to choose parameters a and b which are in accordance with equation 4 and which show good agreement with the measured times t_p of the peak sodium currents and with the inactivation time constants τ_h at all potentials. Results for a rat nerve fibre are given in Fig. 17.

Armstrong (1978) and Armstrong and Croop (1982) have discarded coupled models as shown in scheme 2 because these models predict finite sodium currents during recovery from inactivation which, however, have not been observed under normal conditions. In order to investigate this point, sodium tail currents during a long repolarization period were calculated from scheme 2. Figure 18 illustrates that there is indeed a small and slowly decaying sodium current which persists for several milliseconds after repolarizing to −46 mV.

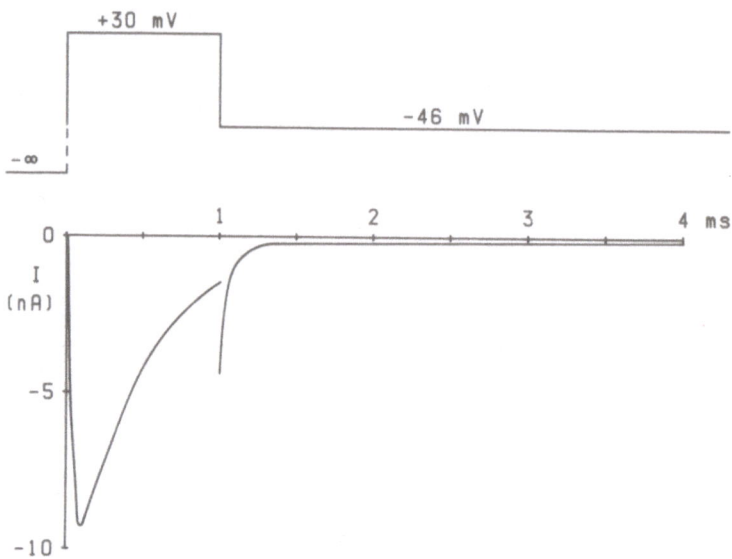

Fig. 18. Simulation of sodium currents I during and after a 1-ms depolarization to $E = 30$ mV. The curves were calculated from scheme 2 with the rate constants (in ms^{-1}): a = 40, b = 1.08, c = 2.3, d = 0.01 at $E = 30$ mV and a = 3.3, b = 6.072, c = 2.3, d = 0.01 at $E = -46$ mV. Probability of state $C_1 = 1$ as initial condition. Currents $I = g \times p_o \times (E - E_{rev})$ with g = 0.3 µS, p_o: probability of open state O and the reversal potential $E_{rev} = 70$ mV

It arises from the transition between the inactivated state I through the open state O into the resting states C_1, C_2 and C_3, but it is so small that it can hardly be distinguished from unspecific and time-independent current components. Hence, the coupled scheme 2 gives an appropriate description of macroscopic sodium currents during and after depolarizations under control conditions.

Problems arise when scheme 2 is used to describe the effects of inactivation-modifying substances. An example is shown in Fig. 15 in which the actions of *Anemonia* toxin II are simulated by varying only the inactivation rates c and d and leaving the activation rates a and b unchanged. With such changes of c and d the slowed and incomplete inactivation can be reproduced, but the calculated peak sodium currents become larger and the predicted time to the peak sodium current becomes longer with ATX II. Since these two changes are in conflict with experiments on frog myelinated nerve (Neumcke et al. 1985), either the coupled scheme 2 does not appropriately describe the effects of ATX II or the toxin not only modifies the process of sodium inactivation but in addition alters sodium activation.

Similar difficulties are encountered when simulating the effects of inactivation-modifying substances on sodium channels in neuroblastoma cells. The large negative voltage shift of the peak sodium conductance curve shown in Fig. 3b can only partly be explained by the coupled scheme 2. This is illus-

Fig. 19. Peak open probabilities p_0 under control conditions (*full curve*) and after complete removal of sodium inactivation (*interrupted curve*). The control p_0 values were calculated from scheme 2 with voltage-independent inactivation rate constants $c = 2.3$ ms^{-1}, $d = 0.01$ ms^{-1} and with activation rate constants a and b taken from Fig. 17a,b. Without sodium inactivation, i.e. for $c = 0$ (*interrupted curve*), the peak probabilities p_0 are equal to their stationary values $(1+b/a)^{-3}$, and b/a is given by equation 4

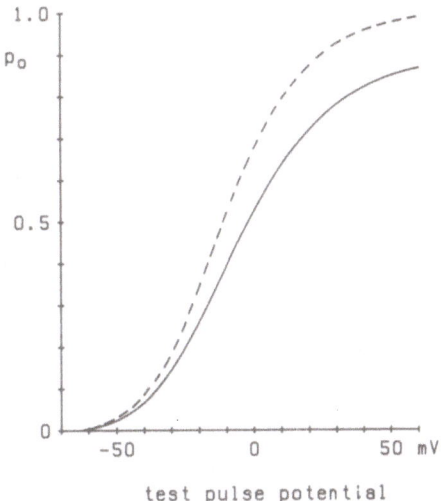

trated in Fig. 19 with plots of the calculated peak "open-probabilities" under control conditions and after removal of sodium inactivation. Thus, the coupled reaction diagram 2 indeed predicts a negative shift of the conductance curve after removal of sodium inactivation, but the shift is much smaller than the one reported by Gonoi and Hille (1987) for N18 neuroblastoma cells.

5.2 Description of Gating Currents

Since the inactivation rate constants c and d in scheme 2 are assumed to be voltage-independent, transitions between the open state O and the inactivated state I occur without charge transfer and thus are electrically silent. Hence, the state I cannot contribute directly to the off-gating current during repolarization, and the reverse charge displacement is delayed because the state O is reached with the slow rate constant d. This explains why the rapid off-charge displacement is smaller than the on-value and why the deficit increases in parallel to sodium inactivation.

While the immobilization of gating charges is readily understandable in terms of the reaction diagram 2, it is less obvious that such a coupled scheme predicts a slow component in the on-gating current. Intuitively, slow current components should not arise from the rapid activation reactions C_1-C_2-C_3-O but only from the slow inactivation step O-I which, however, is electrically silent. Such a separation between fast activation and slow inactivation current components is fulfilled by reaction schemes in which the two gating processes are independent of each other, e.g. in the Hodgkin-Huxley formalism. On the other hand, in scheme 2 with coupled activation and inactivation, all individ-

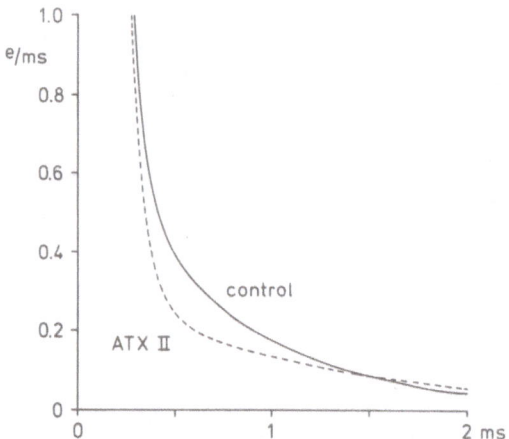

Fig. 20. Simulation of gating currents in frog nerve during a pulse to $E = -10$ mV under control conditions (*full curve*) and after treatment with *Anemonia* toxin ATX II (*interrupted curve*). The currents per channel are given in units of elementary charges e/ms ($= 1.6 \times 10^{-16}$ A) and were calculated from scheme 2 with the same rate constants as in Fig. 17. Charge transfer 1 e for the transitions $C_1 - C_2$, $C_2 - C_3$ and $C_3 - O$, no charge transfer for the transition $O - I$. Probability of state $C_1 = 1$ as initial condition. Initial amplitude of gating current (at 0 ms): 36 e/ms for control and ATX II. Stationary charge displacements: 2.997 e (control), 2.915 e (ATX II). Since the control and ATX II curves are almost identical up to 0.3 ms, only the "foot" of the gating currents for later times is shown

ual transitions are described by fast activation and slow inactivation kinetics. For example, with the rate constants (in ms^{-1}): a = 12, b = 1.8, c = 2.3 and d = 0.01, the time course of the probabilities of all states in scheme 2 is described by four exponentials with the time constants (in µs): 24.15, 36.08, 68.71 and 695.7. Thus the voltage-dependent activation reactions are not only governed by fast kinetics, but they contain additional small components which relax with the largest time constant. In this way slow components in the on-gating currents can arise from the voltage-dependent activation reactions. Figure 20 represents simulations of the on-gating currents from scheme 2 to illustrate such slow components under control conditions and after treatment with *Anemonia* toxin ATX II. It is obvious that the toxin decreases the slow component at the "foot" of the on-gating current, as observed in frog myelinated nerve (see Fig. 10b, c).

5.3 Single Channel Properties

As explained in Sect. 4.3, the rate constants of inactivation and deactivation can be determined separately from measurements of the currents in single sodium channels. In the reaction scheme 2 the rates of the transitions from the open state O are c (inactivation) and 3b (deactivation). The corresponding

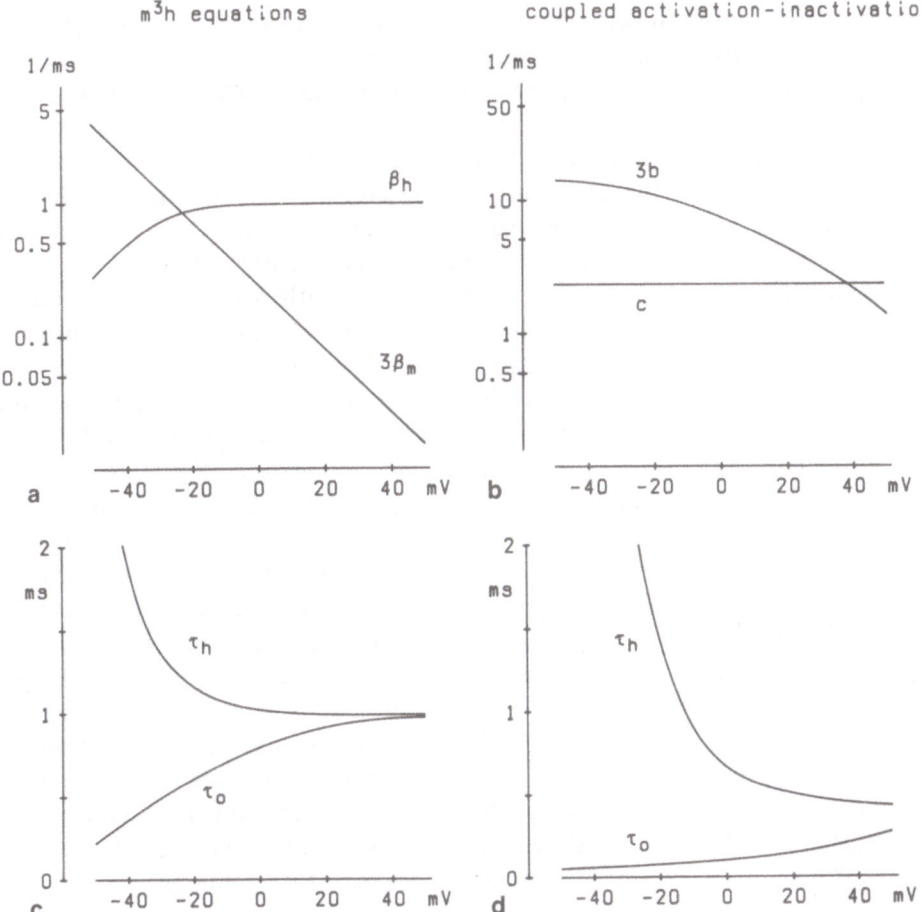

Fig. 21a–d. Rate and time constants of the Hodgkin-Huxley m³h formalism (*left*) and of the reaction scheme 2 with coupled activation, inactivation and with voltage-independent inactivation rates c and d (*right*). **a, b** Voltage dependence of inactivation (β_h, c) and deactivation ($3\,\beta_m$, 3b) rate constants. Note logarithmic scaling of ordinates. Equivalent gating charges (in units of elementary charges *e*) as obtained from the slope of the curves at the membrane potentials V: 1.20 (β_h, $-50\,\text{mV} < V < -30\,\text{mV}$), 0 ($\beta_h$, $V > 0\,\text{mV}$), 1.33 ($3\,\beta_m$), 0.64 (3b near $V = 0\,\text{mV}$), 0 (c). **c, d** Voltage dependence of inactivation time constant (τ_h) and of duration of open sodium channel (τ_0). **a, c** α_h, β_h, β_m, $\tau_h = 1/(\alpha_h + \beta_h)$ and $\tau_0 = 1/(\beta_h + 3\,\beta_m)$ were calculated from equations 23, 24, 21 of Hodgkin and Huxley (1952) assuming a resting potential of $-70\,\text{mV}$. The rates and times refer to sodium currents in squid giant axon at 6.3 °C. **b, d** $\tau_0 = 1/(c + 3\,b)$ was calculated with $c = 2.3\,\text{ms}^{-1}$ and b from Fig. 17b. τ_h values are from Fig. 17d. The rates and times refer to sodium currents in rat myelinated nerve at 20 °C

rates in the Hodgkin-Huxley m³h formalism are β_h for inactivation and $3\,\beta_m$ for deactivation (see Hille 1984 p. 342, Fig. 6). Figure 21 a, b shows the voltage dependencies of the rate constants as calculated from independent m³h or coupled activation and inactivation processes. The two descriptions of sodium currents predict different magnitudes for the rate constants: while with the m^3h formalism inactivation (β_h) is faster than deactivation $(3\beta_m)$ for membrane potentials above -20 mV, the deactivation rate (3 b) predominates over the inactivation rate (c) for potentials up to 40 mV with the coupled scheme 2. These differences explain the different shapes of the τ_h and τ_0 curves in Fig. 21 c, d. With the m^3h gating scheme the duration τ_0 of the open channel approaches the inactivation time constant τ_h at large depolarizations, whereas with coupled activation and inactivation τ_0 remains smaller than τ_h in agreement with experimental results (see Fig. 13).

In conclusion, the coupled reaction scheme 2 with voltage-dependent activation and voltage-independent inactivation rates can predict several properties of macroscopic sodium currents and gating currents which are usually described with the conventional Hodgkin-Huxley formalism. In addition, the coupled scheme can account at least partially for the alteration of sodium activation by inactivation-modifying substances and gives a natural explanation for the immobilization of gating currents owing to sodium inactivation, but it cannot account for the deviations between sodium inactivation and charge immobilization described in Sect. 3.2 (see Armstrong 1978). In addition, some statistical properties of single sodium channels are better described by coupled rather than by independent activation and inactivation processes. As already pointed out, scheme 2 with its strict coupling between activation and inactivation and with its voltage-independent inactivation rates is an extreme case. By relaxing some of the restrictions, i.e. by allowing direct transitions between the inactivated state I and the closed states C_1, C_2 and C_3 or by attributing some voltage dependence to the inactivation rate constants a more realistic scheme of the voltage-dependent sodium channel and a better description of its stationary and kinetic properties could be achieved.

Acknowledgements. I thank Dr. W. Stühmer for communicating unpublished results on sodium channels in *Xenopus* oocytes and Professor B.W. Urban for a fruitful correspondence on sodium channels in lipid bilayers. I am grateful to Professor H. Meves for a critical reading of the manuscript and for many helpful comments. The preparation of this review was supported by the Deutsche Forschungsgemeinschaft (Ne 287/3-2).

References

Adams DJ, Gage PW (1979) Characteristics of sodium and calcium conductance changes produced by membrane depolarization in an *Aplysia* neurone. J Physiol (Lond) 289:143–161

Adelman Jr WJ, Palti Y (1969) The effects of external potassium and long duration voltage conditioning on the amplitude of sodium currents in the giant axon of the squid, *Loligo pealei*. J Gen Physiol 54:589–606

Aldrich RW (1986) Voltage-dependent gating of sodium channels: towards an integrated approach. Trends Neurosci 9:82–86

Aldrich RW, Stevens CF (1983) Inactivation of open and closed sodium channels determined separately. Cold Spring Harbor Symp Quant Biol 48:147–154

Aldrich RW, Stevens CF (1987) Voltage-dependent gating of single sodium channels from mammalian neuroblastoma cells. J Neurosci 7:418–431

Aldrich RW, Corey DP, Stevens CF (1983) A reinterpretation of mammalian sodium channel gating based on single channel recording. Nature 306:436–441

Almers W (1978) Gating currents and charge movements in excitable membranes. Rev Physiol Biochem Pharmacol 82:96–190

Almers W, Stanfield PR, Stühmer W (1983) Slow changes in currents through sodium channels in frog muscle membrane. J Physiol (Lond) 339:253–271

Almers W, Roberts WM, Ruff RL (1984) Voltage clamp of rat and human skeletal muscle: measurements with an improved loose-patch technique. J Physiol (Lond) 347:751–768

Antoni H, Böcker D, Eickhorn R (1988) Sodium current kinetics in intact rat papillary muscle: measurements with the loose-patch-clamp technique. J Physiol (Lond) 406:199–213

Arispe N, Jaimovich E, Liberona JL, Rojas E (1988) Use of selective toxins to separate surface and tubular sodium currents in frog skeletal muscle fibers. Pflügers Arch 411:1–7

Armstrong CM (1978) Models of gating current and sodium conductance inactivation. In: Morad M, Smith S (eds) Biophysical aspects of cardiac muscle. Academic Press, New York

Armstrong CM, Bezanilla F (1974) Charge movement associated with the opening and closing of the activation gates of the Na channels. J Gen Physiol 63:533–552

Armstrong CM, Bezanilla F (1977) Inactivation of the sodium channel. II. Gating current experiments. J Gen Physiol 70:567–590

Armstrong CM, Bezanilla F, Rojas E (1973) Destruction of sodium conductance inactivation in squid axons perfused with pronase. J Gen Physiol 62:375–391

Armstrong CM, Croop RS (1982) Simulation of Na channel inactivation by thiazin dyes. J Gen Physiol 80:641–662

Avenet P, Lindemann B (1987) Patch-clamp study of isolated taste receptor cells of the frog. J Membr Biol 97:223–240

Barchi RL (1983) Protein components of the purified sodium channel from rat skeletal muscle sarcolemma. J Neurochem 40:1377–1385

Barchi RL (1987) Sodium channel diversity: subtle variations on a complex theme. Trends Neurosci 10:221–223

Barnes S, Hille B (1988) Veratridine modifies open sodium channels. J Gen Physiol 91:421–443

Bean BP (1981) Sodium channel inactivation in the crayfish giant axon. Must channels open before inactivating? Biophys J 35:595–614

Bekkers JM, Greeff NG, Keynes RD (1986) The conductance and density of sodium channels in the cut-open squid giant axon. J Physiol (Lond) 377:463–486

Benndorf K, Nilius B (1987) Inactivation of sodium channels in isolated myocardial mouse cells. Eur Biophys J 15:117–127

Benndorf K, Boldt W, Nilius B (1985) Sodium current in single myocardial mouse cells. Pflügers Arch 404:190–196

Benoit E, Dubois JM (1985) Cooperativity of tetrodotoxin action in the frog node of Ranvier. Pflügers Arch 405:237–243

Benoit E, Dubois JM (1987) Interactions of guanidinium ions with sodium channels in frog myelinated nerve fibre. J Physiol (Lond) 391:85–97

Benoit E, Corbier A, Dubois JM (1985) Evidence for two transient sodium currents in the frog node of Ranvier. J Physiol (Lond) 361:339–360

Bergman C, Dubois JM, Rojas E, Rathmayer W (1976) Decreased rate of sodium conductance inactivation in the node of Ranvier induced by a polypeptide toxin from sea anemone. Biochim Biophys Acta 455:173–184

Bezanilla F (1985) Gating of sodium and potassium channels. J Membr Biol 88:97–111

Bezanilla F (1987) Single sodium channels from the squid giant axon. Biophys J 52:1087–1090

Bezanilla F, Armstrong CM (1977) Inactivation of the sodium channel. I. Sodium current experiments. J Gen Physiol 70:549–566

Brismar T (1977) Slow mechanism for sodium permeability inactivation in myelinated nerve fibre of *Xenopus laevis*. J Physiol (Lond) 270:283–297

Brown AM, Lee KS, Powell T (1981) Voltage clamp and internal perfusion of single rat heart muscle cells. J Physiol (Lond) 318:455–477

Bullock JO, Schauf CL (1978) Combined voltage-clamp and dialysis of *Myxicola* axons: behaviour of membrane asymmetry currents. J Physiol (Lond) 278:309–324

Bullock JO, Schauf CL (1979) Immobilization of intramembrane charge in *Myxicola* giant axons. J Physiol (Lond) 286:157–171

Cachelin AB, De Peyer JE, Kokubun S, Reuter H (1983) Sodium channels in cultured cardiac cells. J Physiol (Lond) 340:389–401

Campbell DT (1983) Sodium channel gating currents in frog skeletal muscle. J Gen Physiol 82:679–701

Campbell DT, Hille B (1976) Kinetic and pharmacological properties of the sodium channel of frog skeletal muscle. J Gen Physiol 67:309–323

Carbone E, Lux HD (1986) Na channels in cultured chick dorsal root ganglion neurons. Eur Biophys J 13:259–271

Carmeliet E (1987) Slow inactivation of the sodium current in rabbit cardiac Purkinje fibres. Pflügers Arch 408:18–26

Chandler WK, Meves H (1970a) Evidence for two types of sodium conductance in axons perfused with sodium fluoride solution. J Physiol (Lond) 211:653–678

Chandler WK, Meves H (1970b) Slow changes in membrane permeability and long-lasting action potentials in axons perfused with fluoride solutions. J Physiol (Lond) 211:707–728

Chinn K, Narahashi T (1986) Stabilization of sodium channel states by deltamethrin in mouse neuroblastoma cells. J Physiol (Lond) 380:191–207

Chiu SY (1977) Inactivation of sodium channels: second order kinetics in myelinated nerve. J Physiol (Lond) 273:573–596

Chiu SY (1980) Asymmetry currents in the mammalian myelinated nerve. J Physiol (Lond) 309:499–519

Clark RB, Giles W (1987) Sodium current in single cells from bullfrog atrium: voltage dependence and ion transfer properties. J Physiol (Lond) 391:235–265

Cohen CJ, Bean BP, Colatsky TJ, Tsien RW (1981) Tetrodotoxin block of sodium channels in rabbit Purkinje fibers: interactions between toxin binding and channel gating. J Gen Physiol 78:383–411

Collins CA, Rojas E (1982) Temperature dependence of the sodium channel gating kinetics in the node of Ranvier. Q J Exp Physiol 67:41–55

Collins CA, Rojas E, Suarez-Isla BA (1982a) Activation and inactivation characteristics of the sodium permeability in muscle fibres from *Rana temporaria*. J Physiol (Lond) 324:297–318

Collins CA, Rojas E, Suarez-Isla BA (1982b) Fast charge movements in skeletal muscle fibres from *Rana temporaria*. J Physiol (Lond) 324:319–345

Conti F, Stühmer W (1989) Quantal charge redistributions accompanying the structural transitions of sodium channels. Eur Biophys J 17:53–59

Conti F, DeFelice LJ, Wanke E (1975) Potassium and sodium ion current noise in the membrane of the squid giant axon. J Physiol (Lond) 248:45–82

Conti F, Hille B, Neumcke B, Nonner W, Stämpfli R (1976) Measurement of the conductance of the sodium channel from current fluctuations at the node of Ranvier. J Physiol (Lond) 262:699–727

Conti F, Neumcke B, Nonner W, Stämpfli R (1980) Conductance fluctuations from the inactivation process of sodium channels in myelinated nerve fibres. J Physiol (Lond) 308:217–239

Dodge FA, Frankenhaeuser B (1959) Sodium currents in the myelinated nerve fibre of *Xenopus laevis* investigated with the voltage clamp technique. J Physiol (Lond) 148:188–200 .

Dubois JM, Schneider MF (1982) Kinetics of intramembrane charge movement and sodium current in frog node of Ranvier. J Gen Physiol 79:571–602

Duch DS, Levinson SR (1987) Spontaneous opening at zero membrane potential of sodium channels from eel electroplax reconstituted into lipid vesicles. J Membr Biol 98:57–68

Duch DS, Recio-Pinto E, Frenkel C, Urban BW (1988) Human brain sodium channels in bilayers. Mol Brain Res 4:171–177

Fenwick EM, Marty A, Neher E (1982) Sodium and calcium channels in bovine chromaffin cells. J Physiol (Lond) 331:599–635

Fishman HM (1985) Relaxations, fluctuations and ion transfer across membranes. Prog Biophys Mol Biol 46:127–162

Fox JM (1976) Ultra-slow inactivation of the ionic currents through the membrane of myelinated nerve. Biochim Biophys Acta 426:232–244

Frankenhaeuser B (1960) Quantitative description of sodium currents in myelinated nerve fibres of *Xenopus laevis*. J Physiol (Lond) 151:491–501

Frelin C, Vijverberg HPM, Romey G, Vigne P, Lazdunski M (1984) Different functional states of tetrodotoxin sensitive and tetrodotoxin resistant Na^+ channels occur during the in vitro development of rat skeletal muscle. Pflügers Arch 402:121–128

French RJ, Horn R (1983) Sodium channel gating: models, mimics, and modifiers. Ann Rev Biophys Bioeng 12:319–356

Fuchs W, Hviid Larsen E, Lindemann B (1977) Current-voltage curve of sodium channels and concentration dependence of sodium permeability in frog skin. J Physiol (Lond) 267:137–166

Fujii S, Ayer RK Jr, DeHaan RL (1988) Development of the fast sodium current in early embryonic chick heart cells. J Membr Biol 101:209–223

Fukushima Y (1981) Identification and kinetic properties of the current through a single Na^+ channel. Proc Natl Acad Sci USA 78:1274–1277

Garber SS, Miller C (1987) Single Na^+ channels activated by veratridine and batrachotoxin. J Gen Physiol 89:459–480

Gillespie JI, Meves H (1980) The time course of sodium inactivation in squid giant axons. J Physiol (Lond) 299:289–307

Goldman L (1975) Quantitative description of the sodium conductance of the giant axon of *Myxicola* in terms of a generalized second-order variable. Biophys J 15:119–136

Goldman L, Kenyon JL (1982) Delays in inactivation development and activation kinetics in *Myxicola* giant axons. J Gen Physiol 80:83–102

Goldman L, Schauf CL (1972) Inactivation of the sodium current in *Myxicola* giant axons. Evidence for coupling to the activation process. J Gen Physiol 59:659–675

Goldman L, Schauf CL (1973) Quantitative description of sodium and potassium currents and computed action potentials in *Myxicola* giant axons. J Gen Physiol 61:361–384

Gonoi T, Hille B (1987) Gating of Na channels: inactivation modifiers discriminate among models. J Gen Physiol 89:253–274

Gonoi T, Ohizumi Y, Nakamura H, Kobayashi J, Catterall WA (1987) The *conus* toxin geographutoxin II distinguishes two functional sodium channel subtypes in rat muscle cells developing in vitro. J Neurosci 7:1728–1731

Gordon D, Merrick D, Auld V, Dunn R, Goldin AL, Davidson N, Catterall WA (1987) Tissue-specific expression of the R_I and R_{II} sodium channel subtypes. Proc Natl Acad Sci USA 84:8682–8686

Grant AO, Starmer CF (1987) Mechanisms of closure of cardiac sodium channels in rabbit ventricular myocytes: single channel analysis. Circ Res 60:897–913

Green WN, Weiss LB, Andersen OS (1984) Batrachotoxin-modified sodium channels in lipid bilayers. Ann NY Acad Sci 435:548–550

Green WN, Weiss LB, Andersen OS (1987) Batrachotoxin-modified sodium channels in planar lipid bilayers. Ion permeation and block. J Gen Physiol 89:841–872

Haimovich B, Schotland DL, Fieles WE, Barchi RL (1987) Localization of sodium channel subtypes in adult rat skeletal muscle using channel-specific monoclonal antibodies. J Neurosci 7:2957–2966

Hanke W, Boheim G, Barhanin J, Pauron D, Lazdunski M (1984) Reconstitution of highly purified saxitoxin-sensitive Na^+-channels into planar lipid bilayers. EMBO J 3:509–515

Harris JB, Marshall MW (1973) Tetrodotoxin-resistant action potentials in newborn rat muscle. Nature New Biol 243:191 – 192

Harris JB, Thesleff S (1971) Studies on tetrodotoxin resistant action potentials in denervated skeletal muscle. Acta Physiol Scand 83:382 – 388

Hartshorne RP, Catterall WA (1984) The sodium channel from rat brain: purification and subunit composition. J Biol Chem 259:1667 – 1675

Hille B (1976) Gating in sodium channels of nerve. Ann Rev Physiol 38:139 – 152

Hille B (1984) Ionic channels of excitable membranes. Sinauer Inc, Sunderland Massachusetts USA

Hodgkin AL, Huxley AF (1952) A quantitative description of membrane current and its application to conduction and excitation in nerve. J Physiol (Lond) 117:500 – 544

Hodgkin AL, Huxley AF, Katz B (1952) Measurement of current-voltage relations in the membrane of the giant axon of *Loligo*. J Physiol (Lond) 116:424 – 448

Hodgkin AL, McNaughton PA, Nunn BJ (1985) The ionic selectivity and calcium dependence of the light-sensitive pathway in toad rods. J Physiol (Lond) 358:447 – 468

Horn R, Vandenberg CA (1984) Statistical properties of single sodium channels. J Gen Physiol 84:505 – 534

Horn R, Vandenberg CA (1986) Inactivation of single sodium channels. In: Ritchie JM, Keynes RD, Bolis L (eds) Ion channels in neural membranes. Alan R Liss, New York

Horn R, Patlak J, Stevens CF (1981 a) Sodium channels need not open before they inactivate. Nature 291:426 – 427

Horn R, Patlak J, Stevens CF (1981 b) The effect of tetramethylammonium on single sodium channel currents. Biophys J 36:321 – 327

Hoyt RC (1963) The squid giant axon: mathematical models. Biophys J 3:399 – 431

Hoyt RC (1968) Sodium inactivation in nerve fibers. Biophys J 8:1074 – 1097

Huang LYM, Moran N, Ehrenstein G (1984) Gating kinetics of batrachotoxin-modified sodium channels in neuroblastoma cells determined from single-channel measurements. Biophys J 45:313 – 322

Isenberg G, Ravens U (1984) The effects of the *Anemonia sulcata* toxin (ATX II) on membrane currents of isolated mammalian myocytes. J Physiol (Lond) 357:127 – 149

Jaimovich E, Chicheportiche R, Lombet A, Lazdunski M, Ildefonse M, Rougier O (1983) Differences in the properties of Na^+ channels in muscle surface and T-tubular membranes revealed by tetrodotoxin derivatives. Pflügers Arch 397:1 – 5

Jaimovich E, Ildefonse M, Barhanin J, Rougier O, Lazdunski M (1982) *Centruroides* toxin, a selective blocker of surface Na^+ channels in skeletal muscle: voltage-clamp analysis and biochemical characterization of the receptor. Proc Natl Acad Sci USA 79:3896 – 3900

Jakobsson E (1976) An assessment of a coupled three-state kinetic model for sodium conductance changes. Biophys J 16:291 – 301

Jonas P, Vogel W (1988) Temperature dependence of asymmetry currents in peripheral nerve. Pflügers Arch 411 [Suppl No 1]: R162 (Abstract)

Kayano T, Noda M, Flockerzi V, Takahashi H, Numa S (1988) Primary structure of rat brain sodium channel III deduced from the cDNA sequence. FEBS Lett 228:187 – 194

Keynes RD (1983) Voltage-gated ion channels in the nerve membrane. The Croonian lecture 1983. Proc R Soc Lond [Biol] 220:1 – 30

Keynes RD, Kimura JE (1983) Kinetics of activation of the sodium conductance in the squid giant axon. J Physiol (Lond) 336:621 – 634

Keynes RD, Rojas E (1974) Kinetics and steady-state properties of the charged system controlling sodium conductance in the squid giant axon. J Physiol (Lond) 239:393 – 434

Keynes RD, Rojas E (1976) The temporal and steady-state relationships between activation of the sodium conductance and movement of the gating particles in the squid giant axon. J Physiol (Lond) 255:157 – 189

Keynes RD, Greeff NG, vanHelden DF (1982) The relationship between the inactivating fraction of the asymmetry current and gating of the sodium channel in the squid giant axon. Proc R Soc Lond [Biol] 215:391 – 404

Khodorov BI (1981) Sodium inactivation and drug-induced immobilization of the gating charge in nerve membrane. Prog Biophys Mol Biol 37:49−89

Khodorov BI (1985) Batrachotoxin as a tool to study voltage-sensitive sodium channels of excitable membranes. Prog Biophys Mol Biol 45:57−148

Khodorov BI, Neumcke B, Schwarz W, Stämpfli R (1981) Fluctuation analysis of Na^+ channels modified by batrachotoxin in myelinated nerve. Biochim Biophys Acta 648:93−99

Kimura JE, Meves H (1979) The effect of temperature on the asymmetrical charge movement in squid giant axons. J Physiol (Lond) 289:479−500

Kniffki K-D, Siemen D, Vogel W (1981) Development of sodium permeability inactivation in nodal membranes. J Physiol (Lond) 313:37−48

Kobayashi M, Wu CH, Yoshii M, Narahashi T, Nakamura H, Kobayashi J, Ohizumi Y (1986) Preferential block of skeletal muscle sodium channels by geographutoxin II, a new peptide toxin from *Conus geographus*. Pflügers Arch 407:241−243

Kohlhardt M, Fröbe U, Herzig JW (1987) Properties of normal and non-inactivating single cardiac Na^+ channels. Proc R Soc Lond [Biol] 232:71−93

Kohlhardt M, Fichtner H, Fröbe U (1988) Predominance of poorly reopening single Na^+ channels and lack of slow Na^+ inactivation in neonatal cardiocytes. J Membr Biol 103:283−291

Koppenhöfer E, Schmidt H (1968) Die Wirkung von Skorpiongift auf die Ionenströme des Ranvierschen Schnürrings. I. Die Permeabilitäten P_{Na} and P_K. Pflügers Arch 303:133−149

Krafte DS, Snutch TP, Leonard JP, Davidson N, Lester HA (1988) Evidence for the involvement of more than one mRNA species in controlling the inactivation process of rat and rabbit brain Na channels expressed in *Xenopus* oocytes. J Neurosci 8:2859−2868

Krueger BK, Worley JF III, French RJ (1983) Single sodium channels from rat brain incorporated into planar lipid bilayer membranes. Nature 303:172−175

Krutetskaya ZI, Lonsky AV, Mozhayeva GN, Naumov AP (1978) Two-component nature of the asymmetrical displacement currents in the nerve membrane: the kinetic and pharmacological analysis. Tsitologiya 20:1269−1277 (in Russian)

Kunze DL, Lacerda AE, Wilson DL, Brown AM (1985) Cardiac Na currents and the inactivating, reopening, and waiting properties of single cardiac Na channels. J Gen Physiol 86:691−719

Llano I, Bezanilla F (1984) Analysis of sodium current fluctuations in the cut-open squid giant axon. J Gen Physiol 83:133−142

Meiri H, Spira G, Sammar M, Namir M, Schwartz A, Komoriya A, Kosower EM, Palti Y (1987) Mapping a region associated with Na channel inactivation using antibodies to a synthetic peptide corresponding to a part of the channel. Proc Natl Acad Sci USA 84:5058−5062

Meves H (1974) The effect of holding potential on the asymmetry currents in squid giant axons. J Physiol (Lond) 243:847−867

Meves H (1989) The gating current of the node of Ranvier. In: Narahashi T (ed) Ionic channels II. Plenum Press, New York

Meves H, Nagy K (1989) Multiple conductance states of the sodium channel and of other ion channels. Biochim Biophys Acta 988:99−105

Meves H, Vogel W (1977) Inactivation of the asymmetrical displacement current in giant axons of *Loligo forbesi*. J Physiol (Lond) 267:377−393

Moczydlowski E, Garber SS, Miller C (1984) Batrachotoxin-activated Na^+ channels in planar lipid bilayers. Competition of the tetrodotoxin block by Na^+. J Gen Physiol 84:665−686

Moczydlowski E, Olivera BM, Gray WR, Strichartz GR (1986) Discrimination of muscle and neuronal Na-channel subtypes by binding competition between [^{3}H]saxitoxin and μ-conotoxins. Proc Natl Acad Sci USA 83:5321−5325

Moolenaar WH, Spector I (1978) Ionic currents in cultured mouse neuroblastoma cells under voltage-clamp conditions. J Physiol (Lond) 278:265−286

Nagy K (1987a) Evidence for multiple open states of sodium channels in neuroblastoma cells. J Membr Biol 96:251−262

Nagy K (1987b) Subconductance states of single sodium channels modified by chloramine-T and sea anemone toxin in neuroblastoma cells. Eur Biophys J 15:129–132

Nagy K (1988) Mechanism of inactivation of single sodium channels after modification by chloramine-T, sea anemone toxin and scorpion toxin. J Membr Biol 106:29–40

Nagy K, Kiss T, Hof D (1983) Single Na channels in mouse neuroblastoma cell membrane. Indications for two open states. Pflügers Arch 399:302–308

Nakamura Y, Nakajima S, Grundfest H (1965) The action of tetrodotoxin on electrogenic components of squid giant axons. J Gen Physiol 48:985–996

Narahashi T (1964) Restoration of action potential by anodal polarization in lobster giant axons. J Cell Comp Physiol 64:73–96

Narahashi T, Moore JW, Scott WR (1964) Tetrodotoxin blockage of sodium conductance increase in lobster giant axons. J Gen Physiol 47:965–974

Neumcke B (1982) Fluctuation of Na and K currents in excitable membranes. Int Rev Neurobiol 23:35–67

Neumcke B, Stämpfli R (1982) Sodium currents and sodium-current fluctuations in rat myelinated nerve fibres. J Physiol (Lond) 329:163–184

Neumcke B, Stämpfli R (1983) Alteration of the conductance of Na^+ channels in the nodal membrane of frog nerve by holding potential and tetrodotoxin. Biochim Biophys Acta 727:177–184

Neumcke B, Fox JM, Drouin H, Schwarz W (1976a) Kinetics of the slow variation of peak sodium current in the membrane of myelinated nerve following changes of holding potential or extracellular pH. Biochim Biophys Acta 426:245–257

Neumcke B, Nonner W, Stämpfli R (1976b) Asymmetrical displacement current and its relation with the activation of sodium current in the membrane of frog myelinated nerve. Pflügers Arch 363:193–203

Neumcke B, Nonner W, Stämpfli R (1978) Gating currents in excitable membranes. In: Metcalfe JC (ed) International Review of Biochemistry, Volume 19. University Park Press, Baltimore

Neumcke B, Schwarz W, Stämpfli R (1985) Comparison of the effects of *Anemonia* toxin II on sodium and gating currents in frog myelinated nerve. Biochim Biophys Acta 814:111–119

Neumcke B, Schwarz JR, Stämpfli R (1987) A comparison of sodium currents in rat and frog myelinated nerve: normal and modified sodium inactivation. J Physiol (Lond) 382:175–191

Nilius B (1988) Modal gating behavior of cardiac sodium channels in cell-free membrane patches. Biophys J 53:857–862

Nilius B, Vereecke J, Carmeliet E (1989) Different conductance states of the bursting Na channel in guinea-pig ventricular myocytes. Pflügers Arch 413:242–248

Noda M, Shimizu S, Tanabe T, Takai T, Kayano T, Ikeda T, Takahashi H, Nakayama H, Kanaoka Y, Minamino N, Kangawa K, Matsuo H, Raftery MA, Hirose T, Inayama S, Hayashida H, Miyata T, Numa S (1984) Primary structure of *Electrophorus electricus* sodium channel deduced from cDNA sequence. Nature 312:121–127

Noda M, Ikeda T, Kayano T, Suzuki H, Takeshima H, Kurasaki M, Takahashi H, Numa S (1986a) Existence of distinct sodium channel messenger RNAs in rat brain. Nature 320:188–192

Noda M, Ikeda T, Suzuki H, Takeshima H, Takahashi T, Kuno M, Numa S (1986b) Expression of functional sodium channels from cloned cDNA. Nature 322:826–828

Nonner W (1979) Effects of *Leiurus* scorpion venom on the "gating" current in myelinated nerve. Adv Cytopharmacol 3:345–352

Nonner W (1980) Relations between the inactivation of sodium channels and the immobilization of gating charge in frog myelinated nerve. J Physiol (Lond) 299:573–603

Nonner W, Rojas E, Stämpfli R (1975) Displacement currents in the node of Ranvier. Voltage and time dependence. Pflügers Arch 354:1–18

Nonner W, Rojas E, Stämpfli R (1978) Asymmetrical displacement currents in the membrane of frog myelinated nerve: early time course and effects of membrane potential. Pflügers Arch 375:75–85

Nonner W, Spalding BC, Hille B (1980) Low intracellular pH and chemical agents slow inactivation gating in sodium channels of muscle. Nature 284:360–363

Ochs G, Bromm B, Schwarz JR (1981) A three-state model for inactivation of sodium permeability. Biochim Biophys Acta 645:243–252

Offner FF (1972) The excitable membrane. A physicochemical model. Biophys J 12:1583–1629

Oiki S, Danho W, Montal M (1988) Channel protein engineering: synthetic 22-mer peptide from the primary structure of the voltage-sensitive sodium channel forms ionic channels in lipid bilayers. Proc Natl Acad Sci USA 85:2393–2397

Oxford GS, Pooler JP (1975) Selective modification of sodium channel gating in lobster axon by 2,4,6-trinitrophenol. Evidence for two inactivation mechanisms. J Gen Physiol 66:765–779

Palmer LG (1987) Ion selectivity of epithelial Na channels. J Membr Biol 96:97–106

Pappone PA (1980) Voltage-clamp experiments in normal and denervated mammalian skeletal muscle fibres. J Physiol (Lond) 306:377–410

Parker I, Sumikawa K, Gundersen CB, Miledi R (1988) Expression of Ach-activated channels and sodium channels by messenger RNAs from innervated and denervated muscle. Proc R Soc Lond [Biol] 233:235–246

Patlak JB (1988) Sodium channel subconductance levels measured with a new variance-mean analysis. J Gen Physiol 92:413–430

Patlak JB, Horn R (1982) Effect of N-bromoacetamide on single sodium channel currents in excised membrane patches. J Gen Physiol 79:333–351

Patlak JB, Ortiz M (1985) Slow currents through single sodium channels of the adult rat heart. J Gen Physiol 86:89–104

Patlak JB, Ortiz M (1986) Two modes of gating during late Na^+ channel currents in frog sartorius muscle. J Gen Physiol 87:305–326

Patlak JB, Ortiz M, Horn R (1986) Opentime heterogeneity during bursting of sodium channels in frog skeletal muscle. Biophys J 49:773–777

Peganov EM, Khodorov BI, Shiskova LD (1973) Slow sodium inactivation related to external potassium in the membrane of Ranvier's node. The role of external K. Bull Exp Biol med USSR 25:15–19 (in Russian)

Plant TD (1988) Na^+ currents in cultured mouse pancreatic B-cells. Pflügers Arch 411:429–435

Pröbstle T, Rüdel R, Ruppersberg JP (1988) Hodgkin-Huxley parameters of the sodium channels in human myoballs. Pflügers Arch 412:264–269

Quandt FN (1987) Burst kinetics of sodium channels which lack fast inactivation in mouse neuroblastoma cells. J Physiol (Lond) 392:563–585

Quandt FN, Narahashi T (1982) Modification of single Na^+ channels by batrachotoxin. Proc Natl Acad Sci USA 79:6732–6736

Ravens U (1976) Electromechanical studies of an *Anemonia sulcata* toxin in mammalian cardiac muscle. Naunyn-Schmiedeberg's Arch Pharmacol 296:73–78

Recio-Pinto E, Duch DS, Levinson SR, Urban BW (1987) Purified and unpurified sodium channels from eel electroplax in planar lipid bilayers. J Gen Physiol 90:375–395

Rosenberg RL, Tomiko SA, Agnew WS (1984) Single-channel properties of the reconstituted voltage-regulated Na channel isolated from the electroplax of *Electrophorus electricus*. Proc Natl Acad Sci USA 81:5594–5598

Rudy B (1976) Sodium gating currents in *Myxicola* giant axons. Proc R Soc Lond [Biol] 193:469–475

Ruff RL, Simoncini L, Stühmer W (1987) Comparison between slow sodium channel inactivation in rat slow- and fast-twitch muscle. J Physiol (Lond) 383:339–348

Ruppersberg JP, Rüdel R (1988) Differential effects of halothane on adult and juvenile sodium channels in human muscle. Pflügers Arch 412:17–21

Ruppersberg JP, Schure A, Rüdel R (1987) Inactivation of TTX-sensitive and TTX-insensitive sodium channels of rat myoballs. Neurosci Lett 78:166–170

Sah P, Gibb AJ, Gage PW (1988) The sodium current underlying action potentials in guinea pig hippocampal CA1 neurons. J Gen Physiol 91:373–398

Sakai H, Matsumoto G, Murofushi H (1985) Role of microtubules and axolinin in membrane excitation of the squid giant axon. Adv Biophys 19:43–89

Salkoff L, Butler A, Wei A, Scavarda N, Giffen K, Ifune C, Goodman R, Mandel G (1987) Genomic organization and deduced amino acid sequence of a putative sodium channel gene in *Drosophila*. Science 237:744–749

Scanley BE, Fozzard HA (1987) Low conductance sodium channels in canine cardiac Purkinje cells. Biophys J 52:489–495

Schauf CL, Bullock JO (1979) Modifications of sodium channel gating in *Myxicola* giant axons by deuterium oxide, temperature, and internal cations. Biophys J 27:193–208

Schauf CL, Pencek TL, Davis FA (1976) Slow sodium inactivation in *Myxicola* axons. Evidence for a second inactive state. Biophys J 16:771–778

Schmidt H, Schmitt O (1974) Effect of aconitine on the sodium permeability of the node of Ranvier. Pflügers Arch 349:133–148

Schreibmayer W, Kazerani H, Tritthart HA (1987) A mechanistic interpretation of the action of toxin II from *Anemonia sulcata* on the cardiac sodium channel. Biochim Biophys Acta 901:273–282

Schwarz JR (1986) The effect of temperature on Na currents in rat myelinated nerve fibres. Pflügers Arch 406:397–404

Schwarz W (1979) Temperature experiments on nerve and muscle membranes of frogs. Indications for a phase transition. Pflügers Arch 382:27–34

Sheets MF, Scanley BE, Hanck DA, Makielski JC, Fozzard HA (1987) Open sodium channel properties of single canine cardiac Purkinje cells. Biophys J 52:13–22

Shrager P, Chiu SY, Ritchie JM (1985) Voltage-dependent sodium and potassium channels in mammalian cultured Schwann cells. Proc Natl Acad Sci USA 82:948–952

Sigel E (1987a) Properties of single sodium channels translated by *Xenopus* oocytes after injection with messenger ribonucleic acid. J Physiol (Lond) 386:73–90

Sigel E (1987b) Effects of veratridine on single neuronal sodium channels expressed in *Xenopus* oocytes. Pflügers Arch 410:112–120

Sigworth FJ (1980) The variance of sodium current fluctuations at the node of Ranvier. J Physiol (Lond) 307:97–129

Sigworth FJ, Neher E (1980) Single Na^+ channel currents observed in cultured rat muscle cells. Nature 287:447–449

Simoncini L, Stühmer W (1987) Slow sodium channel inactivation in rat fast-twitch muscle. J Physiol (Lond) 383:327–337

Spitzer NC (1979) Ion channels in development. Annu Rev Neurosci 2:363–397

Starkus JG, Fellmeth BD, Rayner MD (1981) Gating currents in the intact crayfish giant axon. Biophys J 35:521–533

Stevens CF (1986) Analysis of sodium channel function. Prog Zool 33:29–31

Stühmer W, Methfessel C, Sakmann B, Noda M, Numa S (1987) Patch clamp characterization of sodium channels expressed from rat brain cDNA. Eur Biophys J 14:131–138

Stühmer W, Conti F, Suzuki H, Wang X, Noda M, Yahagi N, Kubo H, Numa S (1989) Structural parts involved in activation and inactivation of the sodium channel. Nature 339:597–603

Suzuki H, Beckh S, Kubo H, Yahagi N, Ishida H, Kayano T, Noda M, Numa S (1988) Functional expression of cloned cDNA encoding sodium channel III. FEBS Lett 228:195–200

Swenson Jr RP (1980) Gating charge immobilization and sodium current inactivation in internally perfused crayfish axons. Nature 287:644–645

Swenson Jr RP (1983) A slow component of gating current in crayfish giant axons resembles inactivation charge movement. Biophys J 41:245–249

Taylor RE, Bezanilla F (1983) Sodium and gating current time shifts resulting from changes in initial conditions. J Gen Physiol 81:773–784

Tejedor FJ, Catterall WA (1988) Site of covalent attachment of α-scorpion toxin derivatives in domain I of the sodium channel α subunit. Proc Natl Acad Sci USA 85:8742–8746

Ulbricht W (1977) Ionic channels and gating currents in excitable membranes. Annu Rev Biophys Bioeng 6:7−31

Ulbricht W, Schmidtmayer J (1981) Modification of sodium channels in myelinated nerve by *Anemonia sulcata* toxin II. J Physiol (Paris) 77:1103−1111

Vandenberg CA, Horn R (1984) Inactivation viewed through single sodium channels. J Gen Physiol 84:535−564

Vassilev PM, Scheuer T, Catterall WA (1988) Identification of an intracellular peptide segment involved in sodium channel inactivation. Science 241:1658−1661

Wang GK (1984) Irreversible modification of sodium channel inactivation in toad myelinated nerve fibres by the oxidant chloramine-T. J Physiol (Lond) 346:127−141

Wang GK, Brodwick MS, Eaton DC (1985) Removal of sodium channel inactivation in squid axon by the oxidant chloramine-T. J Gen Physiol 86:289−302

Weiss RE, Horn R (1986) Functional differences between two classes of sodium channels in developing rat skeletal muscle. Science 233:361−364

Yamamoto D, Yeh JZ, Narahashi T (1984) Voltage-dependent calcium block of normal and tetramethrin-modified single sodium channel. Biophys J 45:337−343

Yau K-W, Nakatani K (1984) Cation selectivity of light-sensitive conductance in retinal rods. Nature 309:352−354

Rev. Physiol. Biochem. Pharmacol., Vol. 115
© Springer-Verlag 1990

Potassium Channels in Excitable and Non-excitable Cells

HANS-ALBERT KOLB

Contents

University of Konstanz, Faculty of Biology, D-7750 Konstanz, FRG

1 Introduction

Potassium channels play a crucial role in determining the resting membrane potential, time course, amplitude and polarity of electrical changes in most types of cells. At rest the membrane potential of a typical cell is positive to the potassium equilibrium potential (E_K). When K^+ channels are activated and open, the cell hyperpolarizes and the opening probability of depolarization-dependent calcium and sodium channels is reduced.

With the development of the single channel recording technique by Neher and Sakmann (1976; Hamill et al. 1981), it became possible to study the elementary quantal events within the original cell membrane which underlie the already known, electrophysiologically measured macroscopic currents. With another method for the study of single channel events, ion channels are isolated from the native membrane and reconstituted in model phospholipid bilayers (see Miller 1986; Latorre 1986a, b).

Potassium currents and channels have been described mainly with respect to their properties in excitable cells. Potassium currents have been reviewed by Meech (1978), Lew and Ferreira (1978) and Dubois (1983), while reviews on potassium currents and channel have been presented by Latorre and Miller (1983), Hille (1984), Dubinsky and Oxford (1985), Adams and Galvan (1986), Cook (1988), and Rudy (1988). Some specific reviews have appeared on the role of K^+ channels in modulation of neuronal excitability (Kaczmarek and Levithan 1987), the M-current (Brown 1988a, b), the S-current (Siegelbaum 1987), the ATP-dependent K^+ channel (Noma and Shibasaki 1988; Ashcroft

1988), Ca^{2+}-activated K^+ channels (Lew 1983; Marty 1983a; Schwarz and Passow 1983; Petersen and Maruyama 1984; Ewald et al. 1985) and K^+ channels regulated by nucleotides (Stanfield 1987), biochemical pathways (Duty and Weston 1988), and by guanine nucleotide binding G proteins (Brown and Birnbaumer 1988).

A still increasing number of separate K^+ channels are emerging, and the question arises as to whether there is a common design for all cell membrane channels. For several membrane channels in eukaryotic cells, the amino acid sequences have now been derived by recombinant DNA methods, e.g. the subunits of the nicotinic acetylcholine receptor (Noda et al. 1983), the voltage-dependent sodium channel (Noda et al. 1984), a putative sodium channel in *Drosophila* (Salkoff et al. 1987) and the primary sequence for a gap-junction polypeptide (Paul 1986; Kumar and Gilula 1986). Recently, it has been shown that Shaker messenger ribonucleic acid (mRNA) species can generate functional A-channel types in oocytes (Timpe et al. 1988; see also Sect. 6 of this review). Since the predicted gene products of the Shaker locus show sequence similarities with the voltage-sensitive sodium channel (Noda et al. 1984) and the dihydropyridine receptor of rabbit muscle (Tanabe et al. 1987), it has been suggested that ionic channels are all simply variations of a common structural theme but differ in their gating properties. In the case of K^+ channels, this suggestion is supported by the finding that the sequence of ion selectivity is quite similar in most of the K^+ channels investigated (see Sect. 7.1.2.2).

Three basic types of gating have been identified, according to which the following channel types can be discriminated (Hille 1984):

1. Voltage-sensitive channels, e.g. the classical delayed outward rectifier described by Hodgkin and Huxley, and the inward or anomalously rectifying K^+ channel. While the former is activated by depolarization, the latter is inactivated by depolarization and activated by hyperpolarization (Conti and Neher 1980; Sakmann et al. 1983; Watanabe and Gola 1987).
2. Ion-activated channels which are gated by changes in intracellular (i) ion concentrations. Ca_i^{2+} is capable of activating and Na_i^+ of blocking K^+ channels (Petersen and Maruyama 1984; Marty 1983a, b; Kameyama et al. 1984; Miller et al. 1985; Blatz and Magleby 1986).
3. K^+ channels gated by neurotransmitters like acetylcholine (ACh) (Adams et al. 1982a; Kurachi et al. 1986a). For the muscarinic ACh-operated K^+ channels and the serotonin-sensitive channel, is has been established that G protein binding is involved in channel activation (see Sects. 3.2, 4.2.1).

The gating of K^+ channels may be determined by one or more of these three basic mechanisms (Hille 1984) and can be modulated by phosphorylation (Levitan 1985) and the action of second messengers (Belardetti and Siegelbaum 1988). However, even within one type of channel gating, several pathways of channel activation are possible, as in the case of G protein gating, where several different receptors can activate one K^+ channel population (Kurachi et al. 1986b; Andrade et al. 1986).

Recently, mechanosensitive K^+ channels have been (Guharay and Sachs 1984) identified which might be involved in cellular volume regulation (see Sect. 10). Various types of K^+ channels have also been identified in plant cell membranes, but available data are not yet sufficient to allow classification (reviewed by Hedrich and Schroeder in 1988).

In the past the macroscopic currents have been characterized; the problem now arises of dissecting them and ascertaining the contributions due to the various classes of single channels. The number of authors using the patch-clamp method for identification of single ionic channels is increasing almost exponentially, as is the observed variability of single channel properties, i.e. conductance, the appearance of subconductance states, the ion selectivity, the open probability and the kinetic analysis of dwell-times. The dependence of single channels on agonists, antagonists, cytoplasmic Ca^{2+} and pH, as well as on second messengers, phosphorylation, G protein binding and mechanical stress, has also been investigated. Only in a few excitable cell systems, for example in molluscan ganglion cells, has it been possible to strictly correlate macroscopic currents with the corresponding elementary units − the single channels. In general, though, there exists a mismatch; far more ion channel types have been identified than there are known ion currents. Thus, any macroscopic current must comprise several populations of ion channels.

For example, 11 separate classes of K^+ channels have been characterized in lens cells on the basis of their kinetics and single channel conductance (Rae et al. 1988), and five have been identified in *Helix* ganglia (Ewald et al. 1985). Besides the Ca^{2+}-activated K^+ channel of large conductance, at least four additional classes of voltage-dependent K^+ channels have been identified in phaeochromocytoma cells derived from chromaffin cells of rat adrenal medulla and used as a model system for the regulation of neuronal excitability (Hoshi and Aldrich 1988; Hoshi et al. 1988). It will be difficult to distinguish with certainty their individual contribution to the total macroscopic K^+ current. Are these channels distinct? An alternative interpretation is that there is a single class of K^+ channels, whose gating consists of transitions between different conductance sublevels and kinetic modes (Cull-Candy and Usowicz 1987). However, this interpretation seems to be unlikely, since most of the separate K^+ channel types are generally observed simultaneously in a single membrane patch. In addition, a tissue-specific distribution of K^+ currents and channels should be expected. For example, in heart, ATP sensitive K^+ channels were recorded in pacemaker cells and atrial and ventricular fibres. Apparently this type of K^+ channel appears in every muscle of the heart. By contrast, inward rectifying K^+ channels were observed in ventricular and atrial fibres but not in nodal pacemaker cells, while the muscarinic ACh receptor-operated channel was found in the nodal cell but not in the ventricular fibre.

1.1 Function in Excitable Cells

At least six distinct voltage-dependent K^+ currents can be discriminated in molluscan neurons. Here, the action potential can be used to estimate the total macroscopic K^+ current. The different components of the voltage-dependent K^+ current define the shape and duration, time dependence and firing pattern of action potentials in neurons (Hille 1984) and may be involved in alterations of synaptic efficacy in learning (Shuster et al. 1986; Farley and Auerbach 1986). Since the K^+ currents are mainly inhibitory and responsible for repolarization, an increase in the K^+ conductance depresses electrical excitability.

K^+ currents can be discriminated by their voltage sensitivities, kinetics of activation and inactivation, and pharmacological modulation. Traditionally, they have been classed into a few groups according to type (see Fig. 1 of F. Dreyer in this volume, p. 96, and Kaczmarek and Levithan 1987). Which classes of K^+ channels form the different currents, e.g. which channels underlie the M-current, and how should one proceed to clarify the correlation of macroscopic currents with microscopic channels? There is no general concept available. The selected procedure will certainly depend on the mechanism of channel activation and modulation.

1.2 Function in Non-excitable Cells

In non-excitable systems, the specific physiological function associated with the cellular ion fluxes must be identified and its correlation with different populations of single channels then elucidated. In the following paragraphs, the proposed function of K^+ channels in tissues such as pancreatic B cells (insulin secreting cells), hepatocytes, renal epithelial cells and leucocytes (lymphocytes and macrophages), which have been used for patch-clamp studies, will be briefly summarized.

In pancreatic B cells, the rhythmic electrical activity evoked by glucose is closely involved in the regulation of insulin secretion and depends on variations of the K^+ permeability (Petersen and Findlay 1987).

In hepatocytes, the membrane K^+ permeability determines the rate of voltage-dependent membrane transport processes, such as Na/alanine uptake, and the secretion of bile constituents, such as glutathione and taurocholic acid. It is involved in volume regulatory processes, e.g. after alanine uptake, following disturbances in the acid-base balance or during cell proliferation. K^+

transport may be modulated by hormonal effects (e.g. insulin, glucagon and epinephrine) or sympathetic stimulation.

K^+ secretion and the regulation of K^+ excretion in the kidney are known to occur in the various segments of the nephron (Field and Giebisch 1985). The cortical collecting tubule, a segment of the distal nephron, plays a major role in the regulation of K^+ secretion and Na^+ reabsorption. Another distal segment, the thick ascending limb, functions by reabsorbing salt, but not water, in order to dilute the urine. In both segments, K^+ is pumped into the cell via Na, K adenosine triphosphatase (ATPase) at the basolateral membrane and then is lost through an apical cell membrane conductance that is Ba^{2+} sensitive (Herbert and Andreoli 1984; Cornejo et al. 1987). In addition, K^+ is taken up across the apical membrane of the thick ascending limb by an electrically silent co-transport system with Na and Cl. The overall result is a net secretion of K^+ by the cortical collecting tubule and a net reabsorption of K^+ by the thick ascending limb.

K^+ channels are required for activation by mitogens, increased protein synthesis, interleukin-2 production, thymidine incorporation, and volume regulation (Grinstein et al. 1982) of T lymphocytes, for activation by antigen of allogeneic cells, and for lysis of target cells by natural killer cells (for review see DeCoursey et al. 1985).

1.3 Dissection of K^+ Currents

For simultaneous measurement of both macroscopic and microscopic currents, at least two measuring pipettes (i.e. whole-cell and cell-attached) should be applied to the cell under investigation. This approach has been used quite recently for studying sodium transport (Fahlke et al. 1988). In the case of single cells, the whole-cell current was measured by one pipette, while the single ion channels were simultaneously monitored by a cell-attached pipette. To enable a relevant comparison, both the macroscopic current amplitudes and single channel open probabilities have to be varied simultaneously, e.g. by second messengers (Neher 1988). Another approach involves noise analysis of the fluctuating part of the macroscopic current, but this requires that the current is composed of one or two channel populations exhibiting different kinetics (Conti et al. 1975; Fenwick et al. 1982; DeFelice 1981; Neumcke 1982; Kolb 1984).

Besides correlating single channels with macroscopic currents, it is important to compare ion channels characterized in different tissues. Classification, as in a table of chemical elements, then calls for the selection of appropriate discriminating parameters. At present, the separate channels are ordered solely according to phenomenological features. Even in the case of the Ca^{2+}-dependent K^+ channel of large unit conductance, detailed classification is hindered by the non-comparability of the experimental conditions used by different authors, e.g. the use of different electrolytes or varying concentrations of the blocking cytoplasmic sodium ion. Since this channel type is easily detectable due to its large conductance, it has been intensely studied and will be reviewed in detail later (Sect. 7.1).

2 M-Current: Physiological Function, Distribution and Blockade

The M-current is a small, subthreshold, voltage-dependent outward K^+ current which is activated at more hyperpolarized potentials than the delayed rectifier, the fast Ca^{2+}-activated K^+ current or the A-current (for recent reviews see Brown 1988 a, b). It is the only non-inactivating time- and voltage-sensitive K^+ current, which therefore contributes to the normal resting membrane current. As a background K^+ current it determines the general level of excitability. It becomes activated in the potential region between resting potential and threshold for firing and can therefore limit repetitive activity and the pattern of spike discharges. At least in sympathetic ganglia, the M- and afterhyperpolarization (AHP)-current seem to act additively in the control of excitability.

The M-current was first characterized in cells of frog lumbar sympathetic ganglia (Brown and Adams 1979, 1980; Adams et al. 1982 a, b). Since then it has been identified (for a more complete list, see Brown 1988 a, b) in a variety of different ganglia, cultured spinal cord neurons (Nowak and MacDonald 1983), cortex neurons (Constanti and Galvan 1983), toad stomach smooth muscle cells (Sims et al. 1985) and mouse neuroblastoma cells (Higashida and Brown 1986).

A rich pharmacology is associated with the M-current. It can be inhibited by several different neurotransmitters, cholinergic and peptidergic agonists, Ba^{2+} ($1-4$ mM) (Adams et al. 1982 b; Constanti et al. 1981) and chicken luteinizing hormone-releasing hormone (Jones 1987). The prefix M was chosen because the current is inhibited by muscarinic ACh receptor agonists, including muscarine itself. Evidence suggests an involvement of phosphatidylinositol (PI) turnover and IP_3 in the muscarinic blockade of the M-current in hippocampal pyramidal cells (Dutar and Nicoll 1988). It has also been demonstrated that activators of protein kinase C mimic the muscarinic action in bullfrog sympathetic ganglia (Tsuji et al. 1987). This type of modulation of the M-current is probably responsible for several slow excitatory responses in vertebrate neurons. Tetraethylammonium (TEA) ($1-5$ mM) (Tokimasa 1985), apamin, tubocurarine and hexamethonium do not block this current. The latter three agents are used to separate the M-current from the AHP-current (Pennefather et al. 1985 a, b). So far, the K^+ channels which determine the

M-current have not been identified. This is probably due to their expected long open time, low density (1 μm^{-2}) and estimated low single channel conductance of about 10 pS (Brown 1988a).

3 S-Current

3.1 Inhibitory Effect of Serotonin

Kandel and his colleagues have investigated in detail the gill withdrawal reflex and the mechanism of its sensitization at the level of individual sensory neurons in *Aplysia californica*. The sensitizing stimulus elicits a slow excitatory postsynaptic potential (EPSP) in a facilitatory interneuron. Serotonin (5-HT) has been found to mimic the unknown modulatory transmitter (Kandel and Schwartz 1982). After Klein et al. (1982) stated that the primary effect of 5-HT is a decrease in a specific outward current (S-current), Siegelbaum et al. (1982) characterized a class of 5-HT sensitive K$^+$ channels (the S-channel) in patch-clamp recordings.

5-HT produces prolonged "all-or-none" closure of these channels which results in a slow depolarization. The channels seem to be weakly voltage-dependent and independent of Ca$_i^{2+}$ (Camardo et al. 1983). They are open at resting membrane potential and therefore, like the M-current, contribute to the resting membrane conductance, excitability and repolarization. The channel closure induced by 5-HT results in an increase in the action potential duration and transmitter release from the sensory neuron terminals. The inhibitory effect of 5-HT can largely be mimicked by the catalytic subunit of the cyclic adenosine monophosphate (cAMP)-dependent protein kinase (0.1 – 1 μM) (Camardo et al. 1983; Shuster et al. 1986).

3.2 Role of G Proteins

In contrast, antagonistic modulation of the channel resulting in an increase in the channel open probability is observed after external application of the neuropeptide Phe-Met-Arg-Phe-NH$_2$ (FMRF-amide) (Piomelli et al. 1987; Belardetti and Siegelbaum 1988). FMRF-amide acts through lipoxygenase metabolites of arachidonic acid. Its external application leads to hyperpolarization and a decrease in action potential duration and transmitter release. Since activation of the arachidonic cascade has been associated with the action of G proteins in many vertebrate neurons (reviewed by Axelrod et al. 1988), it was proposed that the S channel is regulated by G proteins independently of the activation of second messengers. This hypothesis was confirmed by intracellular injection of a non-hydrolysable analogue of guanosine tri-phosphate (GTP), guanosine 5'-[γ-thio]trisphosphate (GTP-γS), which mim-icked the hyperpolarizing action of FMRF-amide, but a slower onset was ob-served and its effects were only partially reversible (Sasaki and Sato 1987; Brezina et al. 1987; Piomelli et al. 1987).

Recently, Volterra and Siegelbaum (1988) reported that two different G pro-teins are involved in 5-HT induced cAMP production and FMRF-amide in-itiated release of arachidonic acid. A pertussis toxin-insensitive G protein couples the 5-HT receptor to adenylate cyclase, whereas a pertussis toxin-sen-sitive G protein is proposed as coupling the FMRF-amide receptor to phospholipase A_2, resulting in activation and the release of arachidonic acid. As an alternative, it may be proposed that a G protein signals occupation of the receptor directly to the K^+ channel, inducing it to open. Direct coupling has been reported for the dopamine, histamine and ACh receptors in abdomi-nal ganglion cells of *Aplysia* (Sasaki and Sato 1987), the muscarinic receptor in atrial muscle cells (Pfaffinger et al. 1985; Breitwieser and Szabo 1985; Yatani et al. 1987; Logothetis et al. 1987; see Sect. 4.2.1) and for the 5-HT and gamma-aminobutyric acid (GABA) evoked K^+ current in hippocampal pyramidal cells (Andrade et al. 1986). The appearance of slow EPSPs result-ing from blockade of K^+ channels is common at many levels of vertebrate nervous systems. It is nevertheless not clear whether G proteins and second messengers are involved in all cases.

4 Inward or Anomalously Rectifying K^+ Current

4.1 Mechanism: Role of Intracellular Mg^{2+}

The high conductance of inward (anomalously) rectifying K^+ currents, as opposed to the low conductance of outward K^+ currents, is typical for in-ward rectifying K^+ channels. The inward rectifying K^+ current was first de-scribed by Katz (1949) and later, using the patch-clamp technique on rat myotubes, by Ohmori et al. (1981). Like the M-current and the transient K^+ current, it contributes significantly to the resting membrane potential in verte-brate and invertebrate neurons (Constanti and Galvan 1983; Kandel and Tauc 1966; Llinas 1984; Nelson and Frank 1967; Brew et al. 1986; Stanfield et al. 1985). In egg cells, the increase of the K^+ conductance due to hyperpolariza-tion and decrease due to depolarization mediates action potentials with very long plateaus (Hagiwara 1983).

Armstrong (1966) first suggested a voltage-dependent steric blockade by some intracellular constituents as the mechanism of inward rectification. In the case of the inward K^+ rectifiers mentioned above, intracellular Mg^{2+} is considered to be the blocking agent. The rate constants of the blocking reaction are assumed to be of the same order as the K^+ transport rate through the channel (Vandenberg 1987; Matsuda et al. 1987; for theoretical considerations see Läuger 1985). Hyperpolarization at potentials negative to the E_K causes the inward movement of K^+ and leads to displacement of the blocking ion, and the inward K^+ current then increases; at depolarization, Mg^{2+} blocks the channel by steric hindrance. In the absence of Mg^{2+} the current–voltage relation of the channel becomes linear (Matsuda et al. 1987). A voltage-dependent Mg^{2+} block with remarkably similar properties has also been observed for ATP-sensitive channels (Horie et al. 1987). It has been suggested that the action of Mg^{2+} is superimposed by voltage-dependent conformational changes in the channel protein itself or transmitter binding. These types of channel gating might be modulated by second messengers, as discussed by Stanfield (1988).

4.2 Appearance in Excitable Cells

It is believed that two types of K^+ channels contribute to the inward K^+ current in cardiac ventricular cells (Sakmann et al. 1983): the muscarinic ACh-activated inward K^+ current and an ACh-independent inward rectifying resting K^+ channel. The latter was first identified at this single channel level by Sakmann and Trube (1984a, b). The channel shows complete inward rectification, so that no current flows through this channel at potentials above E_K. With high external K^+ it exhibited interconverting substates with distinct conductances. Interconversion between the closed and the five distinct open states may result from a variable number of protein monomers forming the channel. An approximate square root dependence of the single channel conductance on the external K^+ concentration was found. A density of $0.5 \, \mu m^{-2}$ was estimated for the inward rectifier in pig ventricle cells.

Numerous single channel studies helped to characterize this channel in adult cardiac ventricle cells (Bechem et al. 1983; Kameyama et al. 1983; Trube and Hescheler 1984; Kurachi 1985; Hume and Uehara 1985) and in embryonic chick heart cells (Kell and DeFelice 1988). Depending on the tissue, the channel could be completely blocked by external Ba^{2+} (Sakmann and Trube 1984b; Kameyama et al. 1983), while incomplete closure was reported by Bechem et al. (1983) and Kell and DeFelice (1988) even at $10 \, mM$ $[Ba^{2+}]_o$. Assigning a function to this channel appears to be difficult. It could play a role in countering local elevations in intercellular potassium (Kell and DeFelice 1988).

4.2.1 Mediation by Muscarinic ACh Receptors and the Role of G Proteins

For the second K^+ channel type, which carries the muscarinic (m) ACh receptor modulated inward K^+ current, it has been proposed that a G protein couples the receptor to the channel. Pfaffinger et al. (1985; see also Breitwieser and Szabo 1985) reported that mACh-induced channel activation requires intracellular GTP. The suggestion that a multistep activation process is involved is supported by observations of an intrinsic delay (30 – 100 ms) in the onset of muscarinic activation (Osterrieder et al. 1981).

GTP or non-hydrolysable GTP analogues like GTP-γS dissociate the heterotrimeric G protein into the α-subunit and the $\beta\gamma$ dimer. Attempts to determine which of the G protein subunits are the regulators of the channel activity have led to much controversy. Logothetis et al. (1987) implied the $\beta\gamma$-subunit as active moities, and Codina et al. (1987) the α-subunit. The results of Cerbei et al. (1988) and Kirsch et al. (1988) obtained using atrial cells of embryonic chick, neonatal rat and adult guinea pig, support the involvement of the α-subunit.

Kurachi et al. (1986b) reported that the same population of K^+ channels could be activated by ACh and adenosine in single atrial cells of guinea pig. It was concluded that P1-purinergic and mACh receptors are coupled to the same inward K^+ channel by G proteins, although the receptor density can be different. For example, in rat atrial myocytes the density of adenosine receptors is much lower than that of mACh receptors (Linden et al. 1985). This type of receptor-dependent channel offers a significantly different signal transduction mechanism than does the nicotinic ACh receptor channel. The latter is a stable oligomeric complex, whereas the mACh receptor and the K^+ channel are two distinct molecules. Cytoplasmic second messengers probably do not play a major role in the activation of the mACh dependent K^+ channel. Other inward rectifiers are modulated by transmitters such as 5-HT in molluscan neurons (Sect. 3; Benson and Levitan 1983) and peptides like substance P (Stanfield et al. 1985) and somatostatin (Pennefather et al. 1988).

4.3 Appearance in Non-excitable Cells

Inward rectification has recently become evident in a number of non-excitable cell types, e.g. in several types of renal cells (Gögelein and Greger 1987; Kolb et al. 1987; Hunter et al. 1988b; Parent et al. 1988; Friedrich et al. 1988), pancreatic islet cells (Findlay et al. 1985b) and hepatocytes (Henderson et al. 1988). None of these channels is Ca_i^{2+}-activated in the range of 1 nM to 1 μM Ca^{2+}. The inward single channel conductance in the cell-attached configuration, using a high K^+ electrolyte in the pipette, is about 50 pS in pancreatic islet cells (Cook and Hales 1984; Findlay et al. 1985b), 46 pS in the basolateral membrane of rabbit ascending limb tubule (Gögelein and Greger 1987), 50 pS in the proximal convoluted tubule (Parent et al. 1988) and 44 pS in the hepatocyte (Henderson et al. 1988). In insulin secreting cells it could be blocked by quinine but not by TEA (Findlay et al. 1985c).

It is difficult to attribute a physiological function to this channel in many of these cell types. Inward rectification might help in preventing K^+ loss upon profound depolarization, or it might promote a hyperpolarizing response. However, there are no conditions known under which inward rectification becomes a crucial factor. As mentioned by Parent et al. (1988), inward rectifiers tend to show a more linear conductance under physiological external potassium. In macrophage membranes, hyperpolarization occurs during such functions as phagocytosis, chemotaxis and Fc receptor ligation. Activation of inward rectifying K^+ channels could play a role in restoring the resting potential (cf. Gallin and Sheehy 1985).

5 Delayed Rectifying K^+ Current

5.1 Appearance in Excitable Cells

In squid axon the delayed rectifying K^+ current is the major outward current which is mainly responsible for the repolarizing phase of an action potential. It has been described in detail by Hodgkin and Huxley (1952). In embryonic chick heart ventricle the slope conductance is 15 pS at physiological $[K^+]_o$ (Clapham and Logothetis 1988). No selective pharmacological agonists are known for this current. Like most of the K^+ channel types, it can be blocked in a voltage-dependent manner by TEA^+ (Armstrong 1971) and also by Cs^+ (Wagoner and Oxford 1987). In some cells, such as the neurosecretory bag cell neuron of *Aplysia*, the current may be composed of several kinetically different components. There is evidence that the voltage dependence of activation and inactivation can be modulated by protein phosphorylation (Bezanilla et al. 1985).

5.2 Appearance in Non-excitable Cells

Delayed rectifying outward K^+ channels have been identified in T lymphocytes (Matteson and Deutsch 1984; Cahalan et al. 1985; for review see DeCoursey et al. 1985), B lymphocytes (Choquet et al. 1987), macrophages (Ypey and Clapham 1984), astrocytes (Bevan and Raff 1985), cultured osteoblasts (Ypey et al. 1988), chick hepatocytes (Marchetti et al. 1988) and mammalian platelets (Maruyama 1987). They resemble those in excitable tissues. The single channel slope conductance is about 7 – 10 pS. The channels are not activated by intracellular calcium, at least not in the range of 1 nM to 1 μM. Like their counterparts in nerve and muscle, they can be blocked by the classical K$^+$ channel blockers TEA$^+$, 4-aminopyridine and quinine (Chandy et al. 1984, 1985; DeCoursey et al. 1984). The function of these channels is still unclear. Outward rectifying K$^+$ conductances can be modulated by second messengers (Choquet et al. 1987; Grega et al. 1987) and increased by interleukin 2-stimulated proliferation (Lee et al. 1988), which suggests a role for these channels in non-excitable cells.

6 Transient K$^+$ or A-Current

6.1 Single Channel Properties

The transient K$^+$ current, often called the A-current, differs from the delayed rectifying K$^+$ current in its activation and inactivation kinetics. The A-current activates and inactivates faster during the depolarizing phase. Like the M-current, it is active in the subthreshold region of the membrane potential and can therefore play a role in determining the firing frequency. The single channel slope conductance in *Helix aspera* neurons is 14 pS. It can be blocked by 4-aminopyridine (Taylor 1987). The channel activity can also be modulated by cAMP.

6.2 Expression in Oocytes

On the basis of the electrophysiological analysis of Shaker (a *Drosophila* gene) mutants, the Shaker locus of *Drosophila melanogaster* has been cloned, and it has been proposed that this encodes a structural component of the A channel (Papazian et al. 1987; Kamb et al. 1987). This hypothesis has been confirmed by expression studies in frog oocytes (Timpe et al. 1988). The complementary (c)DNA sequence predicts an integral membrane protein of 70.2 kDa containing several membrane-spanning sequences. It is homologous to the S4 segment present in the vertebrate sodium channel (Noda et al. 1984, 1986) and possibly also in the putative dihydropyridine receptor of the Ca^{2+} channel in skeletal muscle (Tanabe et al. 1987). Since the S4 segment is involved in voltage gating, a conserved mechanism of voltage activation is proposed for the channels mentioned (Tempel et al. 1987). The cDNA-derived amino acid sequence of the presumptive Ca^{2+} channel contains 1873 residues, which are arranged into 24 transmembrane helical stretches in fourfold symmetry. Each of these four distinct motifs contains a helix of about 20 amino acids, the S4 segment, with every third amino acid bearing a positive charge and the intervening two residues being mainly hydrophobic. The structure of the putative *Drosophila* K^+ channel has not yet been elucidated.

The diversity of K^+ channels may be generated by alternatively spliced transcripts (Schwarz et al. 1988). Recently, cDNA clones of mouse brain have been isolated, and their nucleotide sequence predicts a protein similar to the Shaker protein (Tempel et al. 1988). A completely different putative K^+ channel protein of 130 amino acids (cDNA-derived) has been cloned from rat kidney and expressed in *Xenopus* oocytes (Takumi et al. 1988). Since it contains only a single putative transmembrane domain, pore formation by oligomeric structures is proposed. The unusually slow voltage-dependent gating differs from the characteristics of conventional ion channels.

7 Ca^{2+}-Activated K^+ Channels

Two classes of Ca^{2+}-activated K^+ channels have been characterized and are considered to be typical of excitable cells. One is the voltage-dependent K^+ channel of large unit conductance (150 – 300 pS), which is also widespread in non-excitable cells. The single channel properties have been carefully investigated and will be presented here in detail. The second K^+ channel shows little or no voltage dependence and is of small conductance (10 – 14 pS). Both classes can be discriminated by their different sensitivities to the blocking agents charybdotoxin, TEA^+ and apamin. In addition, Ca^{2+}-activated K^+ channels of conductances ranging from 4 pS to 120 pS have been described in excitable and non-excitable cells. Their distinct contribution to the overall cellular K^+ current in the different cell types is less clear.

7.1 Ca^{2+}-Activated K^+ Channels of Large Conductance

7.1.1 Distribution in Various Cell Types

Voltage-sensitive, Ca^{2+}-activated K^+ channels of large conductance have been identified by the patch-clamp technique in nearly every cell type. They were first observed in bovine adrenal chromaffin cell membranes by Marty (1981) and later in cells from skeletal and smooth muscle, ganglia, rat brain synaptosomes and connective tissue. The channel has also been demonstrated in hormone secreting cells, in both apical and basolateral membranes of epithelial cells, and in human macrophages (see Table 1). Latorre et al. (1982) reconstituted the Ca^{2+}-activated K^+ channel from rat muscle in planar lipid bilayers, where it displays properties similar to those observed in the native membrane (Barrett et al. 1982). Various notations are used in the literature

Table 1. Distribution of the Ca^{2+}-activated K^{+}-channel of large unit conductance in various preparations

Preparation	Conductance[a] (pS)	References
Skeletal muscle cells:		
Rat myotubes	187	Pallotta et al. 1981
		McManus and Magleby 1988
Rat myoballs	240	Methfessel and Boheim 1982
Rabbit t-tubule	226	Latorre et al. 1982, 1985
Smooth muscle cells:		
Rabbit longitudinal jejunal	100 – 200	Benham et al. 1986
Rabbit portal vein	273	Inoue et al. 1985, 1986
Frog and toad stomach	200	Berger et al. 1984
Canine airway	266	McCann and Welsh 1986
Heart muscle cells:		
Cow Purkinje fibres	120	Callewaert et al. 1986
Neurons:		
Xenopus spinal	155	Blair and Dionne 1985
Endocrine glands:		
Bovine chromaffin	180	Marty 1981; Marty and Neher 1985
Rat anterior pituitary	208	Wong et al. 1982; Wong and Adler 1986
Exocrine glands:		
Mouse parotid acinar	250	Maruyma et al. 1983a
Mouse submandibular acinar	250	Maruyama et al. 1983a
Human salivary gland	200	Maruyama et al. 1983a
Pig pancreatic acinar	200	Iwatsuki and Petersen 1985; Maruyama et al. 1983b
Pancreatic islet	250	Findlay et al. 1985a
Pancreatic B	244	Cook et al. 1984
Adrenocortical	170	Tabares et al. 1985
Rat lacrimal		Trautmann and Marty 1984
Mouse lacrimal		Findlay 1984
Epithelial cells:		
Choroid plexus	200	Christensen and Zeuthen 1987
Medial cells of human aorta	250	Brezhestovskii et al. 1985
Gallbadder	140	Maruyama et al. 1985
Renal JTC-12.PC3 cells	220	Kolb et al. 1986
Renal proximal tubular cells		Gögelein and Greger 1987
Immune cells:		
Human macrophages	180	Gallin 1984
Synaptosomes:		
Rat brain	200 – 250	Farley and Rudy 1988

[a] Determined by the patch-clamp method with symmetrical high KCl on both sides of the membrane.

to describe this large conductance channel (e.g. BK channel, for Big unitary conductance K^+ channel, or maxi-K_{Ca} channel), but in this review the abbreviation used is maxi-K(Ca) channel.

7.1.2 Single Channel Properties and Channel Blockade

7.1.2.1 Conductance

Maxi-K(Ca) channels have three remarkable properties: (1) large single channel conductance in the range of 150–300 pS, (2) a steep increase in the open probability on depolarization and (3) blockade by TEA^+. The corresponding current–voltage relationship measured in symmetrical $[K^+]$ (140 mM) solutions is nearly linear between -60 mV and $+70$ mV.

The conductance increases with increasing extracellular $[K^+]$. The channel conductance is a non-linear function of $[K^+]_o$. In excised patches the reduction of $[K^+]$ at the extracellular surface to physiological levels results in a decrease in the single channel conductance from about 220 pS to 100 pS (Barrett et al. 1982; Marty and Neher 1985). The functional relationship between conductance and $[K^+]_o$ has been described by the sum of two hyperbolic functions (Moczydlowski et al. 1984; Vergara et al. 1984), which could result from a double occupancy of the channel by K^+, from two channels in parallel, or from a combination of both. Yet, from blockade studies single occupancy of the pore is suggested (see below).

It should be noted that even within one-cell species and 140 mM KCl solutions on both sides of the channel, the conductance can vary within similar ranges, e.g. 190–330 pS (Marty 1981) and 150–240 pS (Barrett et al. 1982). The meaning of this variation in terms of channel structure and ion selectivity has not been considered until now.

Occasionally subconductance states are observed (Barrett et al. 1982; Kolb et al. 1986), but they have neither been included in kinetic schemes nor analysed for channel types with resolved subconductance states (for a review of ion channels with subconductance states see Fox 1987).

7.1.2.2 Ion Selectivity

The maxi-K (Ca) channel is strongly selective for K^+ over Na^+ at physiological K^+ concentrations, as defined by the reversal potential. An upper limit for the Na^+/K^+ permeability ratio of 0.03 has been derived (Yellen 1984). A sequence for its ionic selectivity has been reported in cultured rat muscle cells (Blatz and Magleby 1984):

$$Tl > K > Rb > NH_4 \gg Na, Li$$

This sequence is identical to the one found in the Ca^{2+}-activated K^+ channel of *Aplysia* neurons (Gorman et al. 1982; Hermann and Gorman 1981), and probably holds true for nearly all types of K^+ channels, although variations in the relative permeabilities have been reported. The sequence is the same for the delayed rectifying K^+ current in squid axon (Conti et al. 1975), frog node of Ranvier (Hille 1973) and frog skeletal muscle (Gay and Stanfield 1978). However, the Ca^{2+}-activated K^+ channel from rabbit muscle t-tubule is impermeable to Rb^+ (Latorre and Miller 1983).

For the inward rectifier, the permeability ratios are variable. The relative permeability of NH_4^+/K^+ is considerably lower in inward rectifier channels of starfish eggs (Hagiwara and Takahashi 1974). In snail neurons, the delayed rectifying K^+ channels are more permeable to Cs^+ than to NH_4^+ (Reuter and Stevens 1980). In contrast, the reconstituted voltage-dependent K^+ channel of large conductance (120 pS in 0.1 M K^+; Labarca et al. 1980) from rabbit and frog sarcoplasmic reticulum (SR channel) appears to be less selective. It is also permeable to Na^+, Li^+ and many organic cations (Coronado et al. 1980), but it is blocked by Cs^+ (Coronado and Miller 1979, 1982).

7.1.2.3 Channel Blockade and Structure

Cs^+ reversibly blocks the open maxi-K(Ca) channel in chromaffin cells from either side of the membrane in a voltage-dependent manner (Yellen 1984). The reduction of the current is more pronounced at voltages which should drive Cs^+ into the channel, e.g. external Cs^+ can block inward but not outward K^+ currents and vice versa. The blockade is generally described as a bimolecular chemical reaction (Woodhull 1973) but is too fast to be resolved by kinetic analysis. Therefore, only a reduction in the mean current through the channel can be measured and analysed (Coronado and Miller 1979; Horn and Lange 1983; for a theoretical approach see Läuger 1985).

In the case of other K^+ channels, a result similar to that in chromaffin cells has been reported, i.e. Cs_i^+ blocks the delayed rectifier current (Chandler and Meves 1965; Adelmann and Senft 1966; Bezanilla and Armstrong 1972) and also the SR channel (Coronado and Miller 1979). Yet, the Cs^+ blockade in chromaffin cells is different from the Cs^+ blockade of the t-tubule channel, where Cs^+ blocks only from the outside (Latorre et al. 1985). Ba^{2+} blocks this channel in skeletal muscle by binding within its lumen (Miller et al. 1987). The blocking site is inaccessible to Ba^{2+} when the channel is closed.

Low concentrations (0.1 – 1 mM) of TEA^+ applied to the exterior membrane surface block the Ca^{2+}-activated K^+ channel reversibly (Latorre et al. 1982; Blatz and Magleby 1984; Latorre and Miller 1983; Marty and Neher 1982; Moczydlowski and Latorre 1983; Pallotta et al. 1981). TEA^+ acts from either side of the membrane in rat muscle cells (Blatz and Magleby 1984), chromaffin cells (Yellen 1984) and rabbit muscle t-tubule cells (Vergara and Latorre 1983; Vergara et al. 1984). The voltage dependence of the blocking reaction is much smaller than with Na^+. In both cases, Na^+ (Marty 1983b; Yellen 1984) and TEA^+ decrease the apparent channel conductance. TEA^+ is much more effective on chromaffin cells when externally rather than internally applied. There is evidence that TEA^+ binds to different sites in the Ca^{2+}-activated K^+ channel. The blocking reaction of the blocked and open channel state has been described by a monomolecular chemical reaction. For the corresponding dissociation constant (K_d) following values have been derived: K_d (TEA^+) inside = 30 mM while K_d (TEA^+) outside = 0.2 mM (Yellen 1984). Similar data have been reported by Blatz and Magleby (1984) and Vergara and Latorre (1983) and also for the Ca^{2+}-activated K^+ channels in sympathetic (Adams et al. 1982a) and molluscan neurons (Hermann and Gorman 1981). The properties of these intracellular TEA^+ binding sites are similar to those of the delayed rectifier in giant axons (Armstrong 1971), while the properties of the extracellular site resemble those of the delayed rectifier in the node of Ranvier (Armstrong and Hille 1972).

In mammalian skeletal muscle, nanomolar concentrations of charybdotoxin, derived from the venom of the Isreali scorpion *Leiurus quinquestriatus*, block the active channel through bimolecular interaction (Miller et al. 1985). Quinine, an inhibitor of the Ca^{2+}-induced increase in human red cell K^+ permeability (Armando-Hardy et al. 1975), does not seem to specifically inhibit either the maxi-K(Ca) channel or the Ca^{2+}-activated K^+ channel of about 50 pS in a cultured insulin-secreting cell line (Findlay et al. 1985c).

How can the different maxi-K(Ca) channels be discriminated? They appear to be identical with respect to their voltage sensitivity and Ca_i^{2+} dependence, but distinct in their microscopic kinetics and pharmacology. In cultured myotubes (Blatz and Magleby 1984; Latorre et al. 1982), chromaffin cells (Yellen 1984) and skeletal muscle t-tubule (Vergara and Latorre 1983; Yellen 1984) the channel is blocked by TEA_i, with a K_d of 30 mM or higher (see above). Yet in rat brain (Farley and Rudy 1988) and pituitary cells (Wong and Adler 1986) the channel is blocked by much lower internal TEA^+ concentrations. Considering that the experiments were performed in the presence of various concentrations of internal Na^+, which also reversibly blocks the channel (Marty 1983b), it may still be that the maxi-K(Ca) channels of different tissues share a common structure.

The combination of a large single channel conductance with high K^+ selectivity is intuitively a contradiction. The design of an appropriate selectivity filter for the maxi-K(Ca) channel has been discussed in detail (Latorre and Miller 1983; Schwarz and Passow 1983; Petersen and Maruyama 1984; Latorre et al. 1985; Latorre 1986a,b; for reviews see Yellen 1987 or Gray et al. 1988). The essential geometrical consideration concerning the proposed selectivity filter is mainly based on results obtained with blocking agents (see above). If the entire length of the pore had the same narrow dimensions as the filter part, the conductance would be low. Therefore, Miller (1982; reviewed by Latorre and Miller 1983) proposed a hypothetical structure for the maxi-K(Ca) channel in the sarcoplasmic reticulum of mammalian skeletal muscle. He suggested a pore having a wide entry and exit structure (see also Gray et al. 1988), connected by a narrow, very short tunnel which does not exceed 0.5 nm in length, compared with 5 nm for the entire membrane thickness. The mean cross-sectional area of at least 0.5 nm^2 is reduced to 0.2 nm^2 at the narrowest part of the tunnel, the selectivity region.

Alternatively, Sakmann and Trube (1984a, b) suggested that high-conductance K$^+$ channels are composed of several subunits. For the inward rectifier K$^+$ channel, they attributed the appearance of subconductance levels to the asynchronized openings of the subunits within one channel compound (see also Kell and DeFelice 1988).

It is assumed that the channel contains a maximum of one ion at a time (Blatz and Magleby 1984; for a theoretical approach see Läuger 1973; Hille 1975; Coronado et al. 1980). For K$^+$ channels of small conductance, probably more than one ion occupies the channel simultaneously (for review see Hille and Schwarz 1978).

7.1.2.4 pH Sensitivity

The open probability increases with rising intracellular pH (pancreatic B cells: Cook et al. 1984), the closed time is mainly affected (leaky epithelia: Christensen and Zeuthen 1987). It seems that H$^+$ competes with Ca^{2+} for the same binding site. However, binding of Ca^{2+} leads to channel opening but binding of H$^+$ does not. H$^+$ binding prevents Ca^{2+} from binding to the channel and occurs mainly while the channel is in the closed state. The affinity of the channel to H$^+$ depends on its pK which is about seven. A pH above 7.4 having little additional effect. The pH dependence is similar for the K$^+$ channel of large unit conductance in sarcoplasmic reticulum (Labarca et al. 1980), the inward rectifier in frog skeletal muscle (Blatz 1984) and the K$^+$ conductance of red blood cells (Stampe and Vestergaard-Boginol 1985; Stampe 1985).

The physiological importance of cellular acidification is still under discussion. In pancreatic B cells, a metabolically induced decrease of the intracellular pH may be important in modulating the dynamic pacemaker current to regulate the production of insulin. In skeletal muscle cells, acidification takes place during anaerobic, glycolytic phases which might modulate the Ca^{2+}-dependent K$^+$ conductance (Sahlin et al. 1975).

7.1.3 Kinetic Properties

The large conductance of the channel yields a high signal – noise ratio in the recording, which allows single channel currents to be analysed with high time resolution. Therefore, it is the best kinetically characterized K$^+$ channel.

7.1.3.1 Ca^{2+} and Mg^{2+} Dependence of Kinetic States

The channel kinetics depend only on the Ca^{2+} activity in the cytoplasm, not on extracellular Ca^{2+}. For maxi-K(Ca) channels from rat muscle and epithelial cells, the Ca^{2+}-binding studies are performed either in the excised state of the patched membrane or in reconstitution experiments of the channel in lipid bilayers. Reconstitution experiments with t-tubule vesicles from rabbit and rat skeletal muscle (Latorre et al. 1982; Moczydlowski and Latorre 1983) and smooth muscle plasma membranes from rabbit intestines (Cecchi et al. 1984) in lipid bilayer membranes, demonstrate that Ca^{2+} acts as a ligand (Latorre et al. 1982; Moczydlowski and Latorre 1983; for review see Latorre et al. 1985). Therefore, a Ca^{2+}-dependent biochemical pathway, e.g. a Ca^{2+}-calmodulin interaction or protein kinase C-dependent phosphorylation, is not involved in the regulation of channel opening. However, Ca^{2+}-activated K^+ channels from brain synaptosomes may also be modulated by kinases (Bartschat et al. 1986; Farley and Rudi 1988; Lipkin et al. 1986; Reeves et al. 1986).

The Ca^{2+} sensitivity of the maxi-K(Ca) channel varies considerably from one preparation to another (reviewed by Petersen and Maruyama 1984). Channel activation occurs at micromolar or lower concentrations of Ca^{2+}. Rat muscle cells are maximally sensitive to changes in internal Ca^{2+} between $10^{-6} M$ and $10^{-4} M$ (Barret et al. 1982), a cell line with properties of the proximal tubulus is maximally sensitive between $10^{-7} M$ and $10^{-5} M$ (Kolb et al. 1986) and cells of mammalian exocrine glands are maximally sensitive between $10^{-8} M$ and $10^{-7} M$ (Findlay 1984; Maruyama et al. 1983a, b; Gallacher and Morris 1986). The channel open states are characterized by two, three or even more bound calcium ions (Barrett et al. 1982; Methfessel and Boheim 1982; Magleby and Pallotta 1983a, b; Moczydlowski and Latorre 1983, Vergara and Latorre 1983; Golowasch et al. 1986).

The channel activity may also be regulated by internal Mg^{2+}. At a constant $[Ca^{2+}]_i$ of $10^{-8} M$, Mg^{2+}_i evoked a dose-dependent increase of the channel open probability (Squire and Petersen 1987). The effect is pronounced in the concentration range of $10^{-6} M$ to $10^{-3} M$, which is within the physiological Mg^{2+}_i concentration of 0.4 mM–3.0 mM (Corkey et al. 1986). Maxi-K(Ca) channels from rat skeletal muscle reconstituted in lipid bilayers show a similar Mg^{2+} sensitivity (Golowasch et al. 1986). A physiological function involving fluctuations of the intracellular $[Mg^{2+}]$ has yet to be discovered.

At least two Ca^{2+} ions have to be bound in order to reach a high open probability. Besides the activating effect of Ca^{2+} on the open probability, a decrease of channel conductance with increasing Ca^{2+}_i has been reported by a few authors (Barrett et al. 1982; Maruyama et al. 1983a, b). The microscopic kinetics have been most extensively studied by Latorre and co-workers

(1985) in reconstitution experiments fusing t-tubule vesicles from rabbit or rat skeletal muscle with lipid bilayers and in patch-clamp experiments by Magleby and co-workers (McManus and Magleby 1988) on cultured rat muscle cells.

Four different modes of kinetic activity were discriminated in cultured rat myobubes: normal, intermediate open, brief open and buzz (McManus and Magleby 1988). It is not clear whether the frequency of these modes depends on the Ca^{2+} activity and how it changes in the presence of the blocking intracellular Na^+. The analysis of the dwell-times (Sigworth and Sine 1987) is based on the assumption that the channel enters a finite number of discrete states, with the transition probability between the states remaining constant in time (Cox and Miller 1965; Colquhoun and Hawkes 1981). For the normal mode which covers about 96% of all transitions, application of this Markovian kinetic analysis yields at least three to four open states and at least six to eight shut states, including one or more infrequently adopted long-lived shut states. Shorter open intervals are adjacent to longer shut intervals, indicating that several shut states make direct transitions to corresponding open states (McManus et al. 1985). To discriminate between these states McManus and Magleby (1988) used a record of more than 200000 open and shut times, which was tested to be homogeneously distributed over time. Kinetics of this type lead to multi-exponential distributions of open and closed times (Colquhoun and Hawkes 1977, 1981, 1982; Horn and Lange 1983; Horn 1987). The corresponding characteristic time constants of the dwell-time distributions differ by up to more than five orders of magnitude and range from a few tenths of a millisecond to several seconds. The number of different states, e.g. closed states obtained from fits to the closed-time distributions, depends on the length of current records used for statistical analysis (Magleby and Pallotta 1983a; McManus and Magleby 1988), especially for the distribution of the long waiting times.

A different approach for the analysis of dwell-time distributions of ionic channels has been recently proposed by Liebovitch et al. (1987a, b). The fractal description they used correlates the effective rate constant for the transition between open and closed to the time scale of the measurement. This model describes the channel kinetics by a continuum of states, needing two parameters. McManus and Magleby (1988) have shown that the fractal approach does not correctly describe the closed-time distribution. The model by Läuger (1988) is based on the notion that the transition between the conducting and non-conducting state of the channel creates a transient structural defect within the channel protein which migrates by random walk through the protein matrix in discrete steps. If the defect has returned to the original position the ion pathway is open again. The corresponding random walk model for the defect migration is determined by two parameters, the size of the domain available for defect diffusion and the jumping frequencies of the structural defect within the domain. However, this model has not been applied to the maxi-K(Ca) channel. For the endplate channel (Colquhoun and Sakmann 1983), the model agrees well with the experimental observation of non-single exponential dwell-time distributions.

7.1.3.2 Voltage Sensitivity

The channel open probability has a steep voltage sensitivity. It increases e-fold by depolarizations of $10-15$ mV at all Ca^{2+} concentrations (Methfessel and Boheim 1982; Blatz and Magleby 1984; Moczydlowski and Latorre 1983; Kolb et al. 1986). Moczydlowski and Latorre (1983) discussed the relationship between voltage- and calcium-dependent activation. It has been shown (see above) that one Ca^{2+} ion must bind before the channel can open and that further Ca^{2+} ions can bind to the channel in the open state, yielding an increase of the open probability (see also Begenisich 1988). The binding of calcium to the channel is the reaction step which is voltage-dependent. The ligand itself, not an intrinsic gating charge within the channel structure, introduces the voltage sensitivity. Therefore, the binding constant for calcium should increase with depolarization. This means that an increase in calcium concentration shifts the voltage activation to smaller voltages.

7.1.4 Channel Density

The channel density has not been estimated in excitable cells. In epithelial cells the channel density appears to be rather small. Rat lacrimal gland cells have $50-150$ maxi-K(Ca) channels in the cell membrane (Trautmann and Marty 1984). Assuming a cell diameter of about 15 μm, a channel density of about $0.1-0.2$ μm^{-2} is derived. It has been calculated that about 50 channels are present in one rat salivary acinar cell (Maruyama et al. 1983a, b), which yields a channel density of 0.1 μm^{-2}, assuming a cell diameter of 15 μm. A similar channel density of 0.4 μm^{-2} has been reported in ventricular cells of choroid plexus (Christensen and Zeuthen 1987). Therefore, the maximal contribution, of the open maxi-K(Ca) channel to the membrane permeability is calculated to be of 2.2×10^{-5} cm s^{-1}.

7.2 AHP Channel

7.2.1 Distribution in Excitable Cells

Excitable cells possess a second type of Ca^{2+}-dependent K^+ channel with a conductance of $10-14$ pS (Blatz and Magleby 1986), which has been less thoroughly characterized. It has no apparent voltage dependence and is responsible for a slow (late, up to 1 s) afterhyperpolarization (AHP) following an action potential. Therefore it is called an AHP channel (Pennefather et al. 1985 b). AHP triggers repetitive firing (Barrett and Barrett 1976; Barrett et al. 1981; Kawai and Watanabe 1986). The channel co-exists with the maxi-K(Ca) channel in bullfrog sympathetic ganglion cells (Pennefather et al. 1985 b; Goh and Pennefather 1987) and cultured cells from rat skeletal muscle (Blatz and Magleby 1986). It was identified in outside out membrane patches of primary rat muscle cultures (Romey and Lazdunski 1984) after blocking the maxi-K(Ca) channel with TEA$^+$.

The existence of the AHP channel in non-excitable cell membranes still needs to be demonstrated.

7.2.2 Single Channel Conductance, Kinetics and Ca^{2+} Sensitivity

The single channel current-voltage relationship shows linearity in the applied potential range of -20 mV to -60 mV. For the single channel conductance, $10-14$ pS is derived in symmetrical 140 mM KCl solutions.

At the single channel level, neither the sequence for ion selectivity nor the kinetics has been investigated in detail. The small single channel conductance and the high channel density in membrane patches complicate its separation from other K$^+$ channels. In cultured rat skeletal myotubes, most patches typically contain $10-60$ of these channels, rendering the kinetic analysis very complex. The channels have Ca^{2+} sensitivity approximately ten times higher at negative membrane potentials. They show a half-maximal response with $2-5\times10^{-7}$ M Ca$_i^{2+}$ (Blatz and Magleby 1986). Maxi-K(Ca) and AHP channels cannot be discriminated by their Ca^{2+} sensitivity in depolarized membranes, since maxi-K(Ca) channels show an increase of Ca^{2+} sensitivity with depolarization (Barrett et al. 1982; Methfessel and Boheim 1982; Moczydlowski and Latorre 1983).

7.2.3 Channel Blockade

Externally applied TEA$^+$ at concentrations up to $20-25$ mM has no significant blocking effect on the AHP channel in muscle cells (Romey and Lazdunski 1984; Blatz and Magleby 1986) or on the related AHP-current in sympathetic neurons (Pennefather et al. 1985 b). This fact allows the AHP channel to be distinguished from the maxi-K(Ca) channel.

Nanomolar concentrations of apamin, a peptide from bee venom, can block the AHP channel in these cell systems (Burgess et al. 1981; Romey and Lazdunski 1984; Pennefather et al. 1985 b; Blatz and Magleby 1986; Kawai and Watanabe 1986; Goh and Pennefather 1987) and in cat spinal motorneurons (Zhang and Krnjevic 1987). Apamin-binding sites have been identified in vitro, and may provide the possibility of biochemical isolation of this channel (Hugues et al. 1982; see Glossmann and Striessnig, this volume). In contrast, in guinea pig olfactory cortex neurons no blocking effect by apamin or tubocurarine on the AHP-current was observed, whereas TEA$^+$ reduced this current (Constanti and Sim 1987).

7.3 Ca^{2+}-Dependent K$^+$ Channels of 20−90 pS

Ca^{2+}-dependent K$^+$ channels of intermediate conductance and different Ca^{2+} sensitivity have been observed in a great variety of tissues. Only a few will be mentioned here. Excised membrane patches of *Helix* ganglia contain Ca^{2+}-activated K$^+$ channels of 40−60 pS, which are apparently very similar to those described by Lux et al. (1981) and are modulated by cAMP-dependent protein phosphorylation (Ewald et al. 1985). L-Alanine-mediated activation of Ca^{2+}-activated K$^+$ channels of 30 pS was observed in rat liver cells (Bear and Petersen 1987).

Ca^{2+} activated K$^+$ conductances are present in the apical membranes of cells from several nephron segments. Single channel conductances in the cortical collecting tubule of the rabbit (Hunter et al. 1984, 1986) and the rat (Frindt and Palmer 1987) of about 90 pS, in cultured chick kidney cells of 110 pS (Guggino et al. 1985), in the thick ascending limb of 80 pS (Cornejo et al. 1987), and in *Necturus* proximal tubule of 60 pS (Kawahara et al. 1987) were recorded with a high K$^+$ electrolyte in the pipette. Evidence is available which indicates that forskolin, an activator of adenylate cyclase, and antidiuretic hormone, which acts on distal segments (reviewed by Handler and Orloff 1981), stimulate this Ba^{2+}-sensitive K$^+$ conductance through cAMP-mediated pathways (Herbert and Andreoli 1984). Besides the maxi-K(Ca) channel, two further types of voltage-sensitive and Ca^{2+} dependent K$^+$ channels (75−80 pS and 120−125 pS) have been observed in rat brain synaptosomes. The activity of these channels increased on exposure to the catalytic subunit of cAMP-dependent protein kinase (Farley and Rudy 1988).

8 Physiological Function of Ca^{2+}-Dependent K$^+$ Channels

8.1 Excitable Cells

Since excitability is correlated to the generation of action potentials, the role of the maxi-K(Ca) channel is mainly related to its voltage sensitivity, whereas the Ca^{2+} dependence is thought to fine tune excitability. During the early depolarization of an action potential maxi-K(Ca) channels open, causing fast repolarization of the membrane potential towards the E_K. The channels are turned off rapidly at resting potential. Thus the voltage-dependent entry of

Ca^{2+} is terminated and the excitability, e.g. the firing frequency, returns to resting levels (Pallotta et al. 1981; Romey and Lazdunski 1984).

The pharmacological differences between the maxi-K(Ca) channel and the AHP channel suggest that the attributed macroscopic currents (I_c and I_{AHP}; for a review of macroscopic currents in excitable tissues see Kaczmarek and Levithan 1987 and Fig. 1 of F. Dreyer in this volume) are composed of these two channel populations (Fournier and Crepel 1984; Obaid et al. 1985; Lancaster and Adams 1986; Lancaster et al. 1986; Storm 1987; Lancaster and Nicoll 1988). It is furthermore concluded that the fast AHP current is identical to the TEA^+, charybdotoxin- and voltage-sensitive Ca^{2+}-activated K^+ current, I_c. The block of I_c has no effect on the slow AHP current which is caused by the apamin sensitive K^+ channels.

The contribution of the maxi-K(Ca) channel to the overall cellular current depends not only on intracellular Ca^{2+} and membrane potential, but also on cytoplasmic $[Na^+]$, $[Mg^{2+}]$, pH and external $[K^+]$. The necessary elucidation of the interplay between modulation of these six (at least) parameters and the corresponding cellular function will only be possible if methods are developed which allow simultaneous intracellular monitoring of several parameters.

8.2 Secretory Epithelial Cells

It is proposed that the maxi-K(Ca) channel mainly regulates Ca^{2+} entry in secretory cells (Marty and Neher 1982; Wong et al. 1982; Maruyama et al. 1983 a, b; Petersen and Maruyama 1984; Findlay et al. 1985a) and thereby regulates secretion.

As in excitable cells, the K(Ca) channels in secretory epithelial cells are mainly activated by an increase of cytoplasmic Ca^{2+}. This elevation might be partially caused by activation of voltage-gated Ca^{2+} channels (Douglas 1968; Baker et al. 1971; Baker and Knight 1981; Lux 1983; Reuter 1983), as shown for endocrine cells. Since endocrine cells generally fire action potentials it has been suggested that the physiological function for the maxi-K(Ca) channels in these cells is similar to that in excitable cells. During the depolarization phase, the voltage-gated Ca^{2+} influx opens the Ca^{2+} activated K^+ channels. For example, in adult pancreatic islet B cells and in an insulin-secreting cell line stimulation increases Ca_i^{2+} two- to threefold to about $2-5 \times 10^{-7} M$. Within this range of $[Ca^{2+}]_i$ the maxi-K(Ca) channels become activated (Findlay et al. 1985a). The corresponding K^+ efflux leads to hyperpolarization and thus attenuates further Ca^{2+} influx.

Secretion is under strict nervous and hormonal control in exocrine cells (fluid-secreting non-excitable epithelia, like the exocrine pancreas, gastric mucosa, tracheal epithelium and lacrimal, sweat and salivary glands). Stimulus – secretion coupling in these cells is accompanied by a release of cellular K^+ (Young and van Lennep 1979; Petersen 1980). The release of K^+ is linked to the activation of Ca^{2+}-activated K^+ currents by increased cytoplasmic Ca^{2+}. The contribution of cytoplasmic Ca^{2+} release appears to play a major role in exocrine cells. It has been shown by application of neurotransmitters that an increase in Ca_i^{2+} originates only in part from Ca^{2+} influx (Petersen and Maruyama 1984; Trautmann and Marty 1984; Dubinsky and Oxford 1985).

8.3 Non-secretory Epithelial Cells

Ca^{2+}-activated K^+ channels are present in the apical cell membrane of the proximal tubule, the diluting segment and the cortical collecting tubule of the kidney. In these types of kidney cells, the presence of Ba^{2+}-sensitive K^+-conductance correlates with the presence of Ba^{2+} sensitive Ca^{2+} activated K^+ channels. Increases in cell Ca^{2+}, intracellular pH, or cAMP increase K^+ channel activity and could thus stimulate secretory K^+ flux from the cell into the lumen, down a favourable electrochemical gradient (reviewed by Hunter et al. 1988a).

For most of the epithelial cell systems investigated under normal physiological conditions, it was found in cell-attached measurements that the maxi-K(Ca) channels have an extremely small open probability ($\leq 10^{-7}$) at resting membrane potential (Kolb et al. 1986; Christensen and Zeuthen 1987). Assuming a density of 0.1 μm^{-2}, it then appears that the maxi-K(Ca) channel does not significantly contribute to the resting K^+ current. Even at strong depolarization and high internal Ca^{2+} concentration, the channel is only marginally open and does not, therefore, contribute significantly to the existing K^+ conductance. At present it seems difficult to ascribe a regulatory-function to this channel type in non-secretory epithelia. In contrast, this type of channel has a much higher open probability at physiological values of cytoplasmic Ca^{2+} in secretory cells (see above).

8.4 Leucocytes

The activation of macrophages appears to be related to phagocytosis. Since phagocytosis is related to a rise in cytoplasmic Ca^{2+} (Oliveira-Castro 1983), activation of Ca^{2+} dependent K^+ channels has been proposed and also observed (Gallin 1984; Gallin and Sheehy 1985).

9 ATP-Sensitive K$^+$ Channels

K$^+$ channels sensitive to intracellular ATP occur in heart (Trube and Hescheler 1983, 1984; Noma 1983), pancreatic islet B cells (Cook and Hales 1984; Rorsman and Trube 1985; Findlay et al. 1985b; Misler et al. 1986) and skeletal muscle (Spruce et al. 1985). The slope conductance of the linear portion of the current-voltage relationship is about 35 pS at physiological $[K^+]_o$ (Kakei et al. 1985) and about 50 pS for $[K^+]_o = 145$ mM (Rorsman and Trube 1985). The apparent rectification of the outward K$^+$ current through the channel at depolarization is attributed to the reversible blockade by Mg_i^{2+} and Na_i^+, acting on the inner side of the channel (Horie et al. 1987; Noma and Shibasaki 1988). Such a response to internal Na$^+$ is also known for the maxi-K(Ca) channel (Marty 1983b; Yellen 1984).

These channels open only when the cytoplasmic concentration of ATP falls below a critical level. In excised patches from heart, this level is approximately 0.2 mM. An increase of ATP_i decreases the channel open probability without affecting the single channel current amplitude. From the dose-response curve a Hill coefficient of 3 – 4 was calculated. An increase of ATP affects both the channel open times and closed times. The characteristic open times are shortened, whereas the closed times are increased (Kakei et al. 1985), i.e. channel closure is prolonged. The channel open probability can also be reduced by 5'-adenyl imidodiphosphate in the same concentration range as ATP, by adenosine diphosphate (ADP) and adenosine monophosphate (AMP) at higher concentrations, and by guanosine triphosphate (GTP) and uridine triphosphate (UTP). But ADP either inhibits or stimulates channel opening, depending on whether it is applied in combination with ATP (Findlay 1988). Therefore, the binding site is not specific for the base in the nucleotide, though the nature of the site is not known. Nucleotide regulation of K$^+$ channels has recently been reviewed (Stanfield 1987). The channel density in heart and skeletal muscle cells appears to be quite high. A value of $20 - 60\ \mu m^{-2}$ has been estimated (Spruce et al. 1985). In cardiac cells, K$^+$ channels are inhibited by glycolysis, which is at the same time a source of ATP (Weiss and Lamp 1987). At physiological resting levels of intracellular ATP the channels are closed. Their activation has been implicated as a possible cause of K$^+$ efflux during ischaemia. Opening of K$^+$ channels should counteract increased excitability and shorten the time during which the heart is susceptible to lethal arrhythmias of ischaemic origin.

Insulin release from pancreatic islet B cells is regulated by these channels (Rorsman and Trube 1985; Misler et al. 1986). The cells respond to insulin-releasing concentrations of D-glucose with depolarization. It is suggested that glucose increases cytoplasmic ATP and thereby leads to channel closure. Therefore, the glucose-dependent channel (Ashcroft et al. 1984) is identical to the ATP-dependent channel.

10 Stretch-Activated K⁺ Channels

Ion channels sensitive to membrane tension were first described by Guharay and Sachs (1984) in cultured cells from embryonic chick skeletal muscle, and have since been found in a variety of cells: oocytes, red blood cells, snail heart muscle, smooth muscle cells, lens epithelium, cultured tobacco cells, hair cells, dorsal root ganglion neurons, *Escherichia coli*, vascular endothelium, neuroblastoma cells, yeast, choroid plexus epithelium, and isolated and cultured proximal tubule cells (Kirber et al. 1988). The stretch-activated ion channels identified differ in their ion selectivities. Permeabilities for Na^+ as well as for K^+ and also partially for Ca^{2+} are mostly observed (Christensen 1987; Ubl et al. 1988). So far no correlation between specific ion selectivity and corresponding physiological function has been established.

It is generally assumed that these channels play a crucial role in volume regulation, which is often associated with the regulation of K^+ fluxes. Only a few examples of K^+-selective stretch-activated channels have been reported. In renal proximal tubule, Sackin (1987) observed basolateral Ca^{2+}-independent stretch-activated K^+ channels with a conductance between 26 and 47 pS, depending on the solutions on either side of the membrane. It has been suggested that direct, mechanosensitive coupling may exist between cell volume and basolateral K^+ conductance (Sackin and Palmer 1987). In muscle cells the stretch-activated channels may be the mechanoelectrical transducing elements responsible for the initiation of stretch-induced contraction (Sachs 1986).

11 Conclusion

Due to possible changes of single channel properties during development (Blair and Dionne 1985), inherent kinetic and conductance properties of channels are not be expected. Therefore, single channel conductance and open – close kinetics are inappropriate parameters for classification. Discrimination between the channels by their gating properties and biochemical modulation is also not suitable because of frequent overlap. Sensitivity towards channel blockers may occasionally be a suitable parameter, as in the case of various Ca^{2+}-activated K^+ channels which can be distinguished by their differential response to charybdotoxin or apamin. But as long as the binding sites for the blockers are unknown and may depend on stages of cell differentiation, this approach will be of limited use.

Comparison of K^+ channels in different tissues will yield an open-ended discussion as long as the channels are not cloned and expressed, the channel architecture and single channel properties are not determined, and the ordering parameters are not selected. The genetic approach is in full progress. However, molecular biology can be fully exploited only if the physiological functions of a cell have been detailed, the changes involved in the macroscopic membrane currents characterized, and the underlying single channel populations attributed. This combined approach would yield a valuable classification of the channels with respect to their contribution to the specific function of the cell concerned and would provide a practicable way to improve understanding of the relationship between intercellular and intracellular communication.

*Acknowledgements.*The author whishes to thank Mrs. G. Witz and Mr. R. Somogyi for their support in the technical preparation of the manuscript. The work was supported by the Sonderforschungsbereich 156 of the Deutsche Forschungsgemeinschaft.

References

Adams PR, Galvan M (1986) Voltage-dependent currents of vertebrate neurons and their role in membrane excitability. Adv Neurol 44:137–170

Adams PR, Brown DA, Constanti A (1982a) M-currents and other currents in bullfrog sympathetic neurons. J Physiol (Lond) 330:537–572

Adams PR, Brown DA, Constanti A (1982b) Pharmacological inhibition of the M-current. J Physiol (Lond) 332:223–262

Adelman WJ, Senft JP (1966) Voltage clamp studies on the effect of internal cesium ion on sodium and potassium currents in the squid giant axon. J Gen Physiol 50:279

Andrade R, Malenka RC, Nicoll RA (1986) A G protein couples serotonin and $GABA_B$ receptors to the same channels in hippocampus. Science 235:207–211

Armando-Hardy M, Ellory JC, Ferreira MG, Flemminger S, Lew VL (1975) Inhibition of the calcium-induced increase in the potassium permeability of human red blood cells by quinine. J Physiol (Lond) 250:32P–33P

Armstrong CM (1966) Time course of TEA^+-induced anomalous rectification in squid giant axons. J Gen Physiol 50:491–503

Armstrong CM (1971) Interaction of tetraethylammonium ion derivates with the potassium channels of giant axons. J Gen Physiol 58:413–437

Armstrong CM, Hille B (1972) The inner quaternary ammonium ion receptor in potassium channels of the node of Ranvier. J Gen Physiol 59:388–400

Ashcroft FM (1988) Adenosine 5′-triphosphate-sensitive potassium channels. Annu Rev Neurosci 11:97–118

Ashcroft FM, Harrison DE, Ashcroft SJH (1984) Glucose induces closure of single potassium channels in isolated rat pancreatic beta-cells. Nature 312:446–448

Axelrod J, Burch RM, Jelsema CL (1988) Receptor-mediated activation of phospholipase A_2 via GTP-binding proteins: arachidonic acid and its metabolites as second messengers Trends Neurosci 11:117–123

Baker PF, Knight DE (1981) Calcium control of exocytosis and endocytosis in bovine adrenal medullary cells. Philos Trans R Sod Lond [Biol] 296:83–103

Baker PF, Hodgkin AL, Ridgway EB (1971) Depolarization and calcium entry in squid giant axons. J Physiol (Lond) 218:709–755

Barrett EF, Barrett JN (1976) Separation of two voltage-sensitive potassium currents and demonstration of a tetrodotoxin-resistant calcium current in frog motoneurons. J Physiol (Lond) 255:737–774

Barret JN, Barret EF, Dribin LB (1981) Calcium-dependent slow potassium conductance in rat skeletal myotubes. Dev Biol 82:258−266

Barrett JN, Magleby KL, Pallotta BS (1982) Properties of single calcium-activated potassium channels in cultured rat muscle. J Physiol (Lond) 331:211−230

Bartschat DK, French RJ, Nairn AC, Greengard P, Krueger BK (1986) Cyclic AMP-dependent protein kinase modulation of single, calcium-activated potassium channels from rat brain in planar bilayers. Neurosci Abstr 12:1198

Bear CE, Petersen OH (1987) L-Alanine evokes opening of single Ca^{2+}-activated K^+ channels in rat liver cells. Pflügers Arch 410:342−344

Bechem M, Glitsch HG, Pott L (1983) Properties of an inward rectifying K channel in the membrane of guinea-pig atrial cardio balls. Pflügers Arch 399:186−193

Begenisich T (1988) The role of divalent cations in potassium channels. Trends Neurosci 11:270−273

Benham CD, Bolton TB, Lang RJ, Takewaki T (1986) Calcium-activated potassium channels in single smooth muscle cells of rabbit jejunum and guinea pig mesenteric artery. J Physiol (Lond) 371:45−67

Benson JA, Levitan IB (1983) Serotonin increases an anomalously rectifying K^+ current in the *Aplysia* neuron R15. Proc Natl Acad Sci USA 80:3522−3525

Berger W, Grygorcyk R, Schwarz W (1984) Single K^+ channels in membrane evaginations of smooth muscle cells. Pflügers Arch 402:18−23

Belardetti F, Siegelbaum SA (1988) Up- and down-modulation of single K^+ channel function by distinct second messengers. Trends Neurosci 11:232−238

Bevan S, Raff M (1985) Voltage-dependent potassium currents in cultured astrocytes. Nature 315:229−232

Bezanilla F, Armstrong CM (1972) Negative conductance caused by entry of sodium and cesium ions into the potassium channels of squid axons. J Gen Physiol 60:588−608

Bezanilla F, DiPolo R, Caputo C, Rojas H, Torres ME (1985) K^+ current in squid axon is modulated by ATP. Biophys J 47:222a

Blair LAC, Dionne VE (1985) Developmental acquisition of Ca^{2+}-sensitivity by K^+ channels in spinal neurons. Nature 315:329−331

Blatz AL (1984) Asymmetric proton block of inward rectifier K channels in skeletal muscle. Pflügers Arch 401:402−407

Blatz AL, Magleby KL (1984) Ion conductance and selectivity of single calcium-activated potassium channels in cultured rat muscle. J Gen Physiol 84:1−23

Blatz LA, Magleby KL (1986) Single apamin-blocked Ca-activated K^+ channels of small conductance in cultured rat skeletal muscle. Nature 323:718−720

Breitwieser GE, Szabo G (1985) Uncoupling of cardiac muscarinic and β-adrenergic receptors from ion channels by a guanine nucleotide analogue. Nature 317:538−540

Brew H, Gray PTA, Mobbs P, Attwell D (1986) Endfeet of retinal glial cells have higher densities of ion channels that mediate K^+ buffering. Nature 324:466−468

Brezhestovskii PD, Zamoiskii VL, Serebryakov VN, Toptygin AYU, Anotov AS (1985) The high conductance calcium ion activated potassium channel in the membrane of cultivated smooth muscle cells of the human aortic media. Biol Membr 2:487−498

Brezina V, Eckert R, Erxleben C (1987) Suppression of calcium current by an endogenous neuropeptide in neurons of *Aplysia Californica*. J Physiol (Lond) 388:565−595

Brown AM, Birnbaumer L (1988) Direct G protein gating of ion channels. Am J Physiol 254:H401−H410

Brown DA (1988a) M currents. In: Narahashi T (ed) Ion channels. Plenum, New York, pp 55−94

Brown DA (1988b) M-currents: an update. Trends Neurosci 11:294−299

Brown DA, Adams PR (1979) Muscarinic modification of voltage-sensitive currents in sympathetic neurones. Neurosci Abstr 5:585

Brown DA, Adams PR (1980) Muscarinic suppression of a novel voltage-sensitive K^+-current in a vertebrate neurone. Nature 283:673−676

Burgess CM, Claret M, Jenkinson DH (1981) Effects of quinine and apamin on the Ca-dependent K permeability of mammalian hepatocytes and red cells. J Physiol (Lond) 317:67–90

Cahalan MD, Chandy KG, DeCoursey TE, Gupta S (1985) A voltage-gated potassium channel in human T lymphocytes. J Physiol (Lond) 358:197–237

Callewaert G, Vereecke J, Carmeliet E (1986) Existence of a calcium-dependent potassium channel in the membrane of cow cardiac Purkinje cells. Pflügers Arch 406:424–426

Camardo JS, Shuster MJ, Siegelbaum SA, Kandel ER (1983) Modulation of a specific potassium channel in sensory neurons of *Aplysia* by serotonin and cAMP-dependent protein phosphorylation. Cold Spring Harbor Symp Quant Biol 48:213–220

Cecchi X, Wolff D, Alvarez O, Latorre R (1984) Incorporation of Ca^{2+}-activated K^+ channels from rabbit intestinal smooth muscle sarcolemma, into planar bilayers. Biophys J 45:38a (abstr)

Cerbai E, Klockner U, Isenberg G (1988) The alpha subunit of the GTP-binding protein activates muscarinic potassium channels of the atrium. Science 240:1782–1783

Chandler WK, Meves H (1965) Voltage clamp experiments on internally perfused giant axons. J Physiol (Lond) 180:788–820

Chandy KG, DeCoursey TE, Cahalan MD, McLaughlin C, Gupta S (1984) Voltage-gated potassium channels are required for human T lymphocyte activation. J Exp Med 160:369–385

Chandy KG, DeCoursey TE, Cahalan MD, Gupta S (1985) Ion channels in lymphocytes. J Clin Immunol 5:1–6

Choquet D, Sarthou P, Primi D, Cazenave P, Korn H (1987) Cyclic AMP-modulated potassium channels in murine B cells and their precursors. Science 235:1211–1214

Christensen O (1987) Mediation of cell volume regulation by Ca^{2+}-influx through stretch-activated channels. Nature 330:66–68

Christensen O, Zeuthen T (1987) Maxi K^+ channels in leaky epithelia are regulated by intracellular Ca^{2+}, pH and membrane potential. Pflügers Arch 408:249–259

Clapham DE, Logothetis DE (1988) Delayed rectifier K^+ current in embryonic chick heart ventricle. Am J Physiol 254:H192–H197

Codina J, Yatani A, Grenet D, Brown AM, Birnbaumer L (1987) The alpha subunit of the GTP-binding protein G_K opens atrial potassium channels. Science 236:442–445

Colquhoun D, Hawkes AD (1977) Relaxation and fluctuations of membrane currents that flow through drug-operated channels. Proc R Soc Lond [Biol] 199:231–262

Colquhoun D, Hawkes AD (1981) On the stochastic properties of single ion channels. Proc R Soc Lond [Biol] 211:205–235

Colquhoun D, Hawkes AD (1982) On the stochastic properties of burst of single ion channel openings and of clusters of bursts. Philos Trans R Soc Lond [Biol] 300:1–59

Colquhoun D, Sakmann B (1983) Bursts of openings in transmitter-activated ion channels. In: Sakmann B, Neher E (eds) Single-channel recording. Plenum, New York, pp 345–364

Constanti A, Galvan M (1983) M-currents in voltage-clamped olfactory cortex neurones. Neurosci Lett 39:65–70

Constanti A, Sim JA (1987) Calcium-dependent potassium conductance in guinea-pig olfactory cortex neurones in vitro. J Physiol (Lond) 387:173–194

Constanti A, Adams PR, Brown DA (1981) Why do barium ions imitate acetylcholine? Brain Res 206:244–250

Conti F, Neher E (1980) Single channel recordings of K^+ currents in squid axons. Nature 285:140–143

Conti F, DeFelice LJ, Wanke E (1975) Potassium and sodium ion current noise in the membrane of the squid giant axon. J Physiol (Lond) 248:45–66

Cook DL, Hales CN (1984) Intracellular ATP directly blocks K^+ channels in pancreatic B-cells. Nature 311:271–273

Cook DL, Ikeuchi M, Fujimoto (1984) Lowering of pH_i inhibits Ca^{2+}-activated K^+ channels in pancreatic B-cells. Nature 311:269–273

Cook NS (1988) The pharmacology of potassium channels and their therapeutic potential. Trends Physiol Sci 9:21–28

Corkey BE, Duszynski J, Rich TL, Matschinski B, Williamson JR (1986) Regulation of free and bound magnesium in rat hepatocytes and isolated mitochondria. J Biol Chem 261:2567–2574

Cornejo M, Guggino SE, Guggino WB (1987) Modification of Ca^{2+}-activated K^+ channels in cultured medullary thick ascending limb cells by N-bromoacetamide. J Membr Biol 99:147–155

Coronado R, Miller C (1979) Voltage-dependent caesium blockade of a cation channel from fragmented sarcoplasmic reticulum. Nature 280:807–810

Coronado R, Miller C (1982) Conduction and block by organic cations in a K^+-selective channel from sarcoplasmic reticulum incorporated into planar phospholipid bilayers. J Gen Physiol 79:529–547

Coronado R, Rosenberg RL, Miller C (1980) Ionic selectivity, saturation, and block in a K^+-selective channel from sarcoplasmic reticulum. J Gen Physiol 76:425–446

Cox DR, Miller HD (1965) The theory of stochastic processes. Methuen, Londen

Cull-Candy SG, Usowicz MM (1987) Multiple-conductance channels activated by excitatory amino acids in cerebellar neurons. Nature 325:525–528

DeCoursey TE, Chandy KG, Gupta S, Cahalan MD (1984) Voltage-gated K^+ channels in human T lymphocytes: a role in mitogenesis? Nature 307:465–468

DeCoursey TE, Chandy KG, Gupta S, Cahalan MD (1985) Voltage-dependent ion channels in T-lymphocytes. J Neuroimmunol 10:71–95

DeFelice LJ (1981) Introduction to membrane noise. Plenum, New York

Douglas WW (1968) Stimulus-secretion coupling: the concept and clues from chromaffin and other cells. Br J Pharmacol 34:451–474

Dubinsky JM, Oxford GS (1985) Dual modulation of K channels by thyrotropin-releasing hormone in clonal pituitary cells. Proc Natl Acad Sci USA 82:4282–4286

Dubois JM (1983) Potassium currents in the frog node of Ranvier. Prog Biophys Mol Biol 42:1–20

Dutar P, Nicoll RA (1988) Stimulation of phosphatidylinositol (PI) turnover may mediate the muscarinic suppression of the M-current in hippocampal pyramidal cells. Neurosci Lett 85:89–94

Duty S, Weston AH (1988) The biochemical regulation of potassium channels. Biochem Soc Trans 16:532–534

Ewald DA, Williams A, Levitan IB (1985) Modulation of single Ca^{2+}-dependent K^+-channel activity by protein phosphorylation. Nature 315:503–506

Fahlke CH, Rupperssberg JP, Rüdel R (1988) Simultaneous measurements of the sodium currents in the cell-attached and the whole-cell clamp modes lead to different results. Pflügers Arch 412 [Suppl 1]:22p

Farley J, Auerbach S (1986) Protein kinase C activation induces conductance changes in *Hermissenda* photoreceptors like those seen in associative learning. Nature 319:220–223

Farley J, Rudy B (1988) Multiple types of voltage-dependent Ca-activated K channels of large conductance in rat brain synaptosomal membranes. Biophys J 53:919–934

Fenwick EM, Marty A, Neher E (1982) Sodium channel in bovine chromaffin cells. J Physiol (Lond) 331:599–635

Field MJ, Giebisch GH (1985) Hormonal control of potassium excretion. Kidney Int 27:379–387

Findlay I (1984) A patch-clamp study of potassium channels and whole-cell currents in acinar cells of the mouse lacrimal gland. J Physiol (Lond) 350:179–195

Findlay I (1988) Effects of ADP upon the ATP-sensitive K^+ channel in rat ventricular myocytes. J Membr Biol 101:83–92

Findlay I, Dunne MJ, Peterson OH (1985a) High-conductance K^+ channel in pancreatic islet cells can be activated and inactivated by internal calcium. J Membr Biol 83:169–175

Findlay I, Dunne MJ, Peterson OH (1985b) ATP-sensitive inward rectifier and voltage- and calcium-activated K^+ channels in cultured pancreatic islet cells. J Membr Biol 88:165–172

Findlay I, Dunne MJ, Ullrich S, Wollheim CB, Petersen OH (1985c) Quinine inhibits Ca^{2+}-independent K^+ channels whereas tetramethylammonium inhibits Ca^{2+}-activated K^+ channels in insulin secreting cells. FEBS Lett 185:4—8

Fournier E, Crepel F (1984) Electrophysiological properties of dentate granule cells in mouse hippocampal slices maintained in vitro. Brain Res 311:75—86

Fox JA (1987) Ion channel subconductance states. J Membr Biol 97:1—8

Friedrich F, Paulmichl M, Kolb HA, Lang F (1988) Potassium channels in renal epitheloid cells (MDCK) activated by serotonin. J Membr Biol 106:149—155

Frindt G, Palmer LG (1987) Ca-activated K channels in apical membrane of mammalian CCT, and their role in K secretion. Am J Physiol 252:F458—F467

Gallacher DV, Morris AP (1986) A patch-clamp study of potassium currents in resting and acetylcholine-stimulated mouse submandibular acinar cells. J Physiol (Lond) 373:379—395

Gallin EK (1984) Calcium- and voltage-activated potassium channels in human macrophages. Biophys J 46:821—827

Gallin EK, Sheeby PA (1985) Evidence for both a calcium-activated potassium conductance and an inward rectifying potassium conductance in macrophages. In: van Furth R (ed) Mononuclear phagocytes. Characteristics, physiology and function. Nijhoff, Boston, pp 379—386

Gay LA, Stanfield PR (1978) The selectivity of the delayed potassium conductance of frog skeletal muscle fibers. Pflügers Arch 378:177—179

Gögelein H, Greger R (1987) Properties of single K^+ channels in the basolateral membrane of rabbit proximal straight tubules. Pflügers Arch 410:288—295

Goh JW, Pennefather PS (1987) Pharmacological and physiological properties of the after-hyperpolarization current of bullfrog ganglion neurons. J Physiol (Lond) 394:315—330

Golowasch J, Kirkwood A, Miller C (1986) Allosteric effects of Mg^{2+} on the gating of Ca^{2+}-activated K^+ channels from mammalian skeletal muscle. J Exp Biol 124:5—13

Gorman ALF, Woolum JC, Cornwall MC (1982) Selectivity of Ca-activated and light-dependent K channels for monovalent cations. Biophys J 38:319—322

Gray MA, Tomlins B, Montgomery RAP, Williams AJ (1988) Structural aspects of the sarcoplasmic reticulum K^+ channel revealed by gallamine block. Biophys J 54:233—239

Grega DS, Werz MA, MacDonald RL (1987) Forskolin and phorbol esters reduce the same potassium conductance of mouse neurons in culture. Science 235:345—348

Grinstein S, Dupre A, Rothstein A (1982) Volume regulation by human lymphocytes. J Gen Physiol 79:849—868

Guggino SE, Suarez-Isla BA, Guggino WB, Gree N, Sacktor B (1985) The influence of barium on apical membrane potentials and potassium channel activity in cultured rabbit medullary ascending limb cells (MTAL). Fed Proc 44:443—451

Guharay F, Sachs F (1984) Stretch activated single ion-channel currents in tissue-cultured embryonic chick skeletal muscle. J Physiol (Lond) 352:685—701

Hagiwara S (1983) Membrane potential-dependent ion channels in cell membrane. Raven, New York

Hagiwara S, Takahashi K (1974) The anomalous rectification and cation selectivity of the membrane of a starfish egg cell. J Membr Biol 18:61—80

Hamill OP, Marty A, Neher E, Sakmann B, Sigworth FJ (1981) Improved patch-clamp techniques for high-resolution current recording from cells and cell-free membrane patches. Pflügers Arch 391:85—100

Handler JS, Orloff J (1981) Antidiuretic hormone. Annu Rev Physiol 43:611—624

Hedrich R, Schroeder JI (1989) The physiology of ion channels and electrogenic pumps in higher plants. Annu Rev Plant Physiol 40:539—569

Henderson RM, Graf J, Boyer JL (1988) Inward-rectifying potassium and high conductance anion channels in rat hepatocytes. Am J Physiol

Herbert SC, Andreoli TE (1984) Control of NaCl transport in the thick ascending limb. Am J Physiol 246:F745—F756

Hermann A, Gorman ALF (1981) Effects of tetraethylammonium on potassium currents in a molluscan neuron. J Gen Physiol 78:87–110

Higashida H, Brown DA (1986) Two polyphosphatidylinositide metabolites control two K currents in a neuronal cell. Nature 323:333–335

Hille B (1973) Potassium channels in myelinated nerve: selective permability to small cations. J Gen Physiol 61:669–686

Hille B (1975) Ionic selectivity of Na and K channels of nerve membranes. In: Eisenman G (ed) Membranes, a series of advances. Dekker, New York, pp 255–323

Hille B (1984) Ionic channels of excitable membranes. Sinauer, Sunderland MA, pp 99–116

Hille B, Schwarz W (1978) Potassium channels as multi-ion single-file pores. J Gen Physiol 72:409–442

Hodgkin AL, Huxlay AF (1952) The components of membrane conductance in the squid giant axon of *Logio*. J Physiol (Lond) 116:473–496

Horie M, Irisawa H, Noma A (1987) Voltage-dependent magnesium block of adenosine-triphosphate-sensitive potassium channel in guinea-pig ventricular cells. J Physiol (Lond) 387:251–272

Horn R (1987) Statistical methods for model discrimination. Applications to gating kinetics and permeation of the acetylcholine receptor channel. Biophys J 51:255–263

Horn R, Lange K (1983) Estimating kinetic constants from single channel data. Biophys J 43:207–223

Hoshi T, Aldrich RW (1988) Voltage-dependent K^+ currents and underlying single K^+ channels in pheochromocytoma cells. J Gen Physiol 91:73–106

Hoshi T, Garber SS, Aldrich RW (1988) Effect of forskolin on voltage-gated K^+ channels is independent of adenylate cyclase activation. Science 240:1652–1655

Hugues M, Romey G, Duval D, Vincent JP, Lazdunski M (1982) Apamin as a selective blocker of the Ca-dependent K channel in neuroblastoma cells. Voltage-clamp and biochemical characterization. Proc Natl Acad Sci USA 79:1308–1312

Hume JR, Uehara A (1985) Ionic basis of the different action potential configurations of single guinea-pig atrial and ventricular myocytes. J Physiol (Lond) 366:525–544

Hunter M, Lopes AG, Boulpaep EL, Cohen B (1984) Single channel recordings of calcium-activated potassium channels in the apical membrane of rabbit cortical collecting tubule. Proc Natl Acad Sci USA 81:4237–4239

Hunter M, Lopes AG, Boulpaep EL, Giebisch GH (1986) Regulation of single potassium channels from apical membrane of rabbit collecting tubule. Am J Physiol 251:F725–F733

Hunter M, Kawahara K, Giebisch G (1988a) Calcium-activated epithelial potassium channels. Miner Electrolyte Metab 14:48–57

Hunter M, Oberleithner H, Henderson RM, Giebisch G (1988b) Whole-cell potassium currents in single early distal tubule cells. Am J Physiol 255:F699–F703

Inoue R, Kitamura K, Kuriyama H (1985) Two Ca-dependent K-channels classified by the application of tetraethylammonium distribute to smooth muscle membranes of the rabbit portal vein. Pflügers Arch 405:173–179

Inoue R, Okabe K, Kitamura K, Kuriyama H (1986) A newly identified Ca^{2+}-dependent K^+ channel in the smooth muscle membrane of single cells dispersed from the rabbit portal vein. Pflügers Arch 406:138–143

Iwatsuki N, Petersen OH (1985) Action of tetraethylammonium on calcium-activated potassium channels in pig pancreatic acinar cells studied by patch-clamp single-channel and whole-cell current recording. J Membr Biol 86:139–144

Jones SW (1987) Chicken II luteinizing hormone-releasing hormone inhibits the M-current of bullfrog sympathetic neurons. Neurosci Lett 80:180–184

Kaczmarek LK, Levithan (1987) Neuromodulation: the biochemical control of neuronal excitability. Oxford University Press, New York

Kakei M, Noma A, Shibasaki T (1985) Properties of adenosine triphosphate-regulated channels in guinea-pig ventricular cells. J Physiol (Lond) 363:441–462

Kamb A, Iverson LE, Tanouye MA (1987) Molecular characterization of Shaker, a *Drosophila* gene that encodes a potassium channel. Cell 50:405–413

Kameyama M, Kiyosue T, Soejima M (1983) Single current analysis of the inward rectifier K current in the rabbit ventricular cells. Jpn J Physiol 33:1039–1056

Kameyama M, Kakei M, Sato T, Shibasaki T, Matsuda H, Irisawa H (1984) Intracellular Na$^+$ channel in mammalian cardiac cells. Nature 309:354–356

Kandel ER, Schwartz JH (1982) Molecular biology of learning: modulation of transmitter release. Science 218:433–443

Kandel ER, Tauc L (1966) Anomalous rectification in the metacerebral giant cells and its consequences for synaptic transmission. J Physiol (Lond) 183:287–304

Katz B (1949) Les constants électriques de la membrane du muscle. Arch Sci Physiol 3:285–299

Kawahara K, Hunter M, Giebisch GH (1987) Potassium channels in *Necturus* proximal tubule. Am J Physiol 253:F488–F494

Kawai T, Watanabe M (1986) Blockade of Ca-activated K conductance by apamin in rat sympathetic neurons. Br J Pharmacol 87:225–232

Kell MJ, DeFelice LJ (1988) Surface charge near the cardiac inward-rectifier channel measured from single channel conductances. J Membr Biol 102:1–10

Kirber MT, Walsh JV jr, Singer JJ (1988) Stretch-activated ion channels in smooth muscle: a mechanism for the initiation of stretch-induced contraction. Pflügers Arch 412:339–345

Kirsch GE, Yatani A, Codina J, Birnbaumer L, Brown AM (1988) Alpha-subunit of G_K activates atrial K$^+$ channels of chick, rat, and guinea pig. Am J Physiol 254:H1200–H1205

Klein M, Camardo J, Kandel ER (1982) Serotonin modulates a specific potassium current in the sensory neurons that show presynaptic facilitation in *Aplysia*. Proc Natl Acad Sci USA 79:5713–5717

Kolb HA (1984) Measuring the properties of single channels in cell membranes. In: Stein WD (ed) Current topics in membranes and transport, vol 21. Academic, New York, pp 133–179

Kolb HA, Brown CDA, Murer H (1986) Characterization of a Ca-dependent maxi K channel in the apical membrane of a cultured renal epithelium (JTC-12.p3). J Membr Biol 92:207–215

Kolb HA, Paulmichl M, Lang F (1987) Epinephrine activates outward rectifying K channel in Madin-Darby canine kidney cells. Pflügers Arch 408:584–591

Kumar NM, Gilula NB (1986) Cloning and characterization of human and rat liver cDNAs coding for a gap junction protein. J Cell Biol 103: 767–776

Kurachi Y (1985) Voltage-dependent activation of the inward rectifier potassium channel in the ventricular cell membrane of guinea-pig heart. J Physiol (Lond) 366:365–385

Kurachi Y, Nakajima T, Sugimoto T (1986a) Acetylcholine activation of K$^+$ channels in cell-free membrane of atrial cells. Am J Physiol 251:H681–H684

Kurachi Y, Nakajima T, Sugimoto T (1986b) On the mechanism of activation of muscarinic K$^+$ channels by adenosine in isolated atrial cells: involvement of GTP-binding proteins. Pflügers Arch 407:264–274

Labarca P, Coronado R, Miller C (1980) Thermodynamic and kinetic studies of the gating behavior of a K$^+$-selective channel from the sarcoplasmic reticulum membrane. J Gen Physiol 76:397–424

Lancaster B, Adams PR (1986) Calcium-dependent current generating the afterhyperpolarization of hippocampal neurons. J Neurophysiol 55:1268–1282

Lancaster B, Nicoll RA (1988) Properties of two calcium-activated hyperpolarizations in rat hippocampal neurones. J Physiol (Lond) 389:187–203

Lancaster B, Madison DV, Nicoll RA (1986) Charybdotoxin selectively blocks a fast Ca-dependent afterhyperpolarization (AHP) in hippocampal pyramidal cells. Neurosci Abstr 12:560

Latorre R (1986a) Ionic channels in cells and model systems. Plenum, New York

Latorre R (1986b) In: Miller C (ed) ion channel reconstitution. Plenum, New York, pp 431–467

Latorre R, Miller C (1983) Conduction and selectivity in potassium channels. J Membr Biol 71:11–30

Latorre R, Vergara C, Hidalgo C (1982) Reconstitution in planar bilayers of Ca^{2+}-dependent potassium channel from transverse tubule membranes isolated from rabbit skeletal muscle. Proc Natl Acad Sci USA 79:805–809

Latorre R, Alvarez O, Cecchi X, Vergara C (1985) Properties of reconstituted ion channels. Annu Rev Biophys Biophys Chem 14:79–111

Läuger P (1973) Ion transport through pores: a rate-theory analysis. Biochim Biophys Acta 311:423–441

Läuger P (1985) Ionic channels with conformational substates. Biophys J 47:581–591

Läuger P (1988) Internal motions in proteins and gating kinetics of ionic channels. Biophys J 53:877–884

Lee SC, Sabath DE, Deutsch C, Prystowsky MB (1988) Increased voltage-gated potassium conductance during interleukin 2-stimulated proliferation of a mouse helper T lymphocyte clone. J Cell Biol 102:1200–1208

Levithan IB (1985) Phosphorylation of ion channels. J Membr Biol 87:177–190

Lew VL (1983) (ed) Ca^{2+}-activated K^+ channels: collected papers and reviews. Cell Calcium 4:321–518

Lew LV, Ferreira HG (1978) Calcium transport and the properties of a calcium-activated potassium channel in red cell membranes. In: Kleint-Zeller A, Bronner F (eds) Current topics in membranes and transport, vol 10. Academic, New York, pp 217–277

Liebovitch LS, Fischbarg J, Koniarek JP (1987a) Ion channel kinetics: a model based on fractal scaling rather than multistate Markov processes. Math Biosci 84:37–68

Liebovitch LS, Fischbarg J, Koniarek JP, Todorova I, Wang M (1987b) Fractal model of ion-channel kinetics. Biochim Biophys Acta 896:173–180

Linden J, Hollen CE, Patel A (1985) The mechanism by which adenosine and cholinergic agents reduce contractility in rat myocardium. Circ Res 56:728–735

Lipkin S, Farley J, Rudy B (1986) Protein kinase C effects on single K channels from mammalian brain. Soc Neurosci Abstr 13:1343

Llinas R (1984) Comparative electrobiology of mammalian central neurons. In: Dingledine R (ed) Brain slices. Plenum, New York, pp 7–24

Logothetis DE, Kurachi Y, Galper J, Neer EJ, Clapham DE (1987) The beta gamma subunits of GTP-binding proteins activate the muscarinic K^+ channel in heart. Nature 325:321–326

Lux HD (1983) Observations on single calcium channels – an overview. In: Sakmann BM, Neher E (eds) Single-channel recording. Plenum, New York, pp 437–449

Lux HD, Neher E, Marty A (1981) Single channel activity associated with the calcium dependent outward current in *Helix pomatia*. Pflügers Arch 389:293–295

Magleby KL, Pallotta BS (1983a) Calcium-dependence of open and shut interval distributions from calcium-activated potassium channels in cultured rat muscle. J Physiol (Lond) 344:585–604

Magleby KL, Pallotta BS (1983b) Burst kinetics of single calcium-activated potassium channels in cultured rat muscle. J Physiol (Lond) 344:605–623

Marchetti C, Premont RT, Brown AM (1988) A whole-cell and single-channel study of the voltage-dependent outward potassium current in avian hepatocytes. J Gen Physiol 91:255–274

Marty A (1981) Calcium-dependent channels with large unitary conductance in chromaffin cell membranes. Nature 291:497–500

Marty A (1983a) Ca^{2+}-dependent K^+ channels with large unitary conductance. Trends Neurosci 6:262–265

Marty A (1983b) Blocking of large unitary calcium-dependent potassium currents by internal sodium ions. Pflügers Arch 396:179–181

Marty A, Neher E (1982) Ionic channels in cultured rat pancreatic islet cells. J Physiol (Lond) 326:36P–37P

Marty A, Neher E (1985) Potassium channels in cultured bovine adrenal chromaffin cells. J Physiol (Lond) 367:117–141

Maruyama Y (1987) A patch-clamp study of mammalian platelets and their voltage-gated potassium current. J Physiol (Lond) 391:467–485

Maruyama Y, Gallagher DV, Peterson OH (1983a) Voltage and Ca-activated K channel in basolateral acinar cell membranes of mammalian salivary glands. Nature 302:827–829

Maruyama Y, Petersen OH, Flanagan P, Pearson GT (1983b) Quantification of Ca^{2+}-activated K^+ channels under hormonal control in pig pancreas acinar cells. Nature 305:228–232

Maruyama Y, Moore D, Petersen OH (1985) Calcium-activated cation channel in rat thyroid follicular cells. Biochim Biophys Acta 821:229–232

Matsuda H, Saigusa A, Irisawa H (1987) Ohmic conductance through the inwardly rectifying K channel and blocking by internal Mg^{2+}. Nature 325:156–159

Matteson DR, Deutsch C (1984) K channels in T lymphocytes: a patch clamp study using monoclonal antibody adhesion. Nature 307:468–471

McCann JD, Welsh MJ (1986) Calcium-activated potassium channels in canine airway smooth muscle. J Physiol (Lond) 372:113–127

McManus OB, Magleby KL (1988) Kinetic states and modes of single large-conductance calcium-activated potassium channels in cultured rat skeletal muscle. J Physiol (Lond) 402:79–120

McManus OB, Blatz AL, Magleby KL (1985) Inverse relationship of the durations of adjacent open and shut intervals for Cl and K channels. Nature 317:625–627

Meech RW (1978) Calcium-dependent potassium activation in nervous tissues. Annu Rev Biophys Bioeng 7:1–18

Methfessel C, Boheim G (1982) The gating of single calcium-dependent potassium channels is described by an activation/blockade mechanism. Biophys Struct Mech 9:35–60

Miller C (1982) Bis-quaternary ammonium blockers as structural probes of the sarcoplasmic reticulum K^+ channel. J Gen Physiol 79:869–891

Miller C (1986) Ion channel reconstitution. Plenum, New York

Miller C, Moczydlowski E, Latorre R, Phillips M (1985) Charybdotoxin, a protein inhibitor of single Ca^{2+}-activated K^+ channels from mammalian skeletal muscle. Nature 313:316–318

Miller C, Latorre R, Reisin I (1987) Coupling of voltage-dependent gating and Ba^{2+} block in the high-conductance, Ca^{2+}-activated K^+ channel. J Gen Physiol 90:427–449

Misler S, Falke LC, Gillis K, McDaniel (1986) A metabolite-regulated potassium channel in rat pancreatic B cells. Proc Natl Acad Sci USA 83:7119–7123

Moczydlowski E, Latorre R (1983) Gating kinetics of Ca^{2+}-activated potassium channels from rat muscle incorporated into planar lipid bilayers: evidence for two voltage-dependent Ca^{2+} binding reactions. J Gen Physiol 82:511–542

Moczydlowski E, Hall S, Garber SS, Strichartz GS, Miller C (1984) Voltage-dependent blockade of muscle Na^+ channels by *Guanidinium* toxins. J Gen Physiol 84:687–704

Neher E (1988) The use of the patch clamp technique to study second messenger-mediated cellular events. Neuroscience 26:727–734

Neher E, Sakmann B (1976) Single-channel currents recorded from membrane at denervated frog muscle membrane. Nature 260:799–802

Nelson PG, Frank K (1967) Anomalous rectification in cat spinal motoneurons and effect of polarizing currents on excitatory postsynaptic potential. J Neurophysiol 30:1097–1113

Neumcke B (1982) Fluctuation of Na and K currents in excitable membranes. Int Rev Neurobiol 23:35–67

Noda M, Furutani Y, Takahashi H, Toyosato M, Tanabe T, Shimizu S, Kikyotani S, Kayano T, Hirose T, Inayama S, Numa S (1983) Cloning and sequence analysis of calf cDNA and human genomic DNA encoding α-subunit precursor of muscle acetylcholine receptor. Nature 305:818–823

Noda M, Shimizu S, Tanabe T, Takai T, Kayano T, Ikeda T, Takahashi H, Nakayama Y, Minamino N, Kangawa K, Matsuo H, Raftery MA, Hirose T, Inayama S, Hayashida H, Miyata T, Numa S (1984) Primary structure of *Electrophorus electricus* sodium channel deduced from cDNA sequence. Nature 312:121–127

Noda M, Ikeda T, Kayano T, Suzuki H, Takeshima H, Kurasaki M, Takahashi H, Numa S (1986) Existence of distinct sodium channel messenger RNAs in rat brain. Nature 320:188–192

Noma A (1983) ATP-regulated K channels in cardiac muscle. Nature 305:147–148

Noma A, Shibasaki T (1988) Intracellular ATP and cardiac membrane currents. In: Narahashi T (ed) Ion channels. Plenum, New York

Nowak LM, MacDonald RL (1983) Muscarine-sensitive voltage-dependent potassium current in cultured murine spinal cord neurones. Neurosci Lett 35:85−91

Obaid AL, Langer D, Salzberg BM (1985) Charybdotoxin (CTX) selectively blocks a calcium-mediated potassium conductance that contributes to the action potential recorded optically from nerve terminals of the frog neurohypophysis. Soc Neurosci Abstr 11:789

Ohmori H, Yoshida S, Hagiwara S (1981) Single K channel currents of anomalous rectification in cultured rat myotubes. Proc Natl Acad Sci USA 78:4960−4964

Oliveira-Castro GM (1983) Ca^{2+}-sensitive K^+ channels in phagocytic cell membranes. Cell Calcium 4:475−492

Osterrieder W, Yang QF, Trautwein W (1981) The time course of the muscarinic response to ionophoretic acetylcholine application to the S-A node of the rabbit heart. Pflügers Arch 389:283−291

Pallotta BS, Magleby KL, Barrett JN (1981) Single channel recordings of a Ca^{2+}-activated K^+ current in rat muscle cell culture. Nature 293:471−474

Papazian DM, Schwarz TL, Tempel BL, Jan YN, Jan LY (1987) Cloning of genomic and complementary DNA from Shaker, a putative potassium channel gene from *Drosophila*. Science 237:749−753

Parent L, Cardinal J, Sauve R (1988) Single-channel analysis of a K channel at basolateral membrane of rabbit proximal convoluted tubule. Am J Physiol 254:F105−F113

Paul DL (1986) Molecular cloning of cDNA for rat liver gap junction protein. J Cell Biol 103:123−134

Pennefather P, Jones SW, Acad PR (1985a) Modulation of repetitive firing in bullfrog sympathetic ganglion cells by two distinct K currents. Neurosci Abstr 11:148

Pennefather P, Lancaster B, Adams PR, Nicoll RA (1985b) Two distinct Ca-dependent K currents in bullfrog sympathetic ganglion cells. Proc Natl Adam Sci USA 82:3040−3044

Pennefather PS, Heisler S, MacDonald JF (1988) A potassium conductance contributes to the action of somatostatin-14 to suppress ACTH secretion. Brain Res 444:346−350

Petersen OH (ed) (1980) The electrophysiology of gland cells. Academic, London

Petersen OH, Findlay I (1987) Electrophysiology of the pancreas. Physiol Rev 67:1054−1116

Petersen OH, Maruyama Y (1984) Calcium-activated potassium channels and their role in secretion. Nature 307:693−696

Pfaffinger PJ, Martin JM, Hunter DD, Nathanson NM, Hille B (1985) GTP-binding proteins couple cardiac muscarinic receptors to a K channel. Nature 317:536—538

Piomelli D, Shapiro E, Feinmark SJ, Schwartz JH (1987) Metabolites of arachidonic acid in the nervous system of *Aplysia*: possible mediators of synaptic modulation. J Neurosci 7:3675−3686

Rae JL, Levis RA, Eisenberg RS (1988) Ionic channels in ocular epithelia. In: Narahashi T (ed) Ion channels. Plenum, New York, pp 283−327

Reeves R, Farley J, Rudy B (1986) cAMP dependent protein kinase opens several K channels from mammalian brain. Soc Neurosci Abstr 13:1343

Reuter H (1983) Calcium channels and their modulation by neurotransmitters, enzymes and drugs. Nature 301:569−574

Reuter H, Stevens CF (1980) Ion conductance and ion selectivity of potassium channels in snail neurons. J Membr Biol 57:103−118

Romey G, Lazdunski M (1984) The coexistence in rat muscle cells of two distinct classes of Ca^{2+}-dependent K^+ channels with different pharmacological properties and different physiological functions. Biochem Biophys Res Commun 118:669−674

Rorsman P, Trube G (1985) Glucose dependent K^+-channels in pancreatic β-cells are regulated by intracellular ATP. Pflügers Arch 405:305−309

Rudy B (1988) Diversity and ubiquity of K channels. Neuroscience 25:729−749

Sachs F (1986) Biophysics of mechanoreception. Membr Biochem 6:173−195

Sackin H (1987) Stretch-activated potassium channels in renal proximal tubule. Am J Physiol 253:F1253–F1262

Sackin H, Palmer LG (1987) Basolateral potassium channels in renal proximal tubule. Am J Physiol 253:F476–487

Sahlin K, Harris RC, Hultman E (1975) Creatine kinase equilibrium and lactate content compared with muscle pH in tissue samples obtained after isometric exercise. Biochem J 152:173–180

Sakmann B, Trube G (1984a) Conductance properties of single inwardly rectifying potassium channels in ventricular cells from guinea-pig heart. J Physiol (Lond) 347:641–657

Sakmann B, Trube G (1984b) Voltage-dependent inactivation of inward-rectifying single channel currents in the guinea-pig heart cell. J Physiol (Lond) 347:659–683

Sakmann B, Noma A, Trautwein W (1983) Acetylcholine activation of single muscarinic K^+ channels in isolated pacemaker cells of the mammalian heart. Nature 303:250–253

Salkoff L, Butler A, Wei A, Scavarda N, Giffen K, Ifune C, Goodman R, Mandel G (1987) Genomic organization and deduced amino acid sequence of a putative sodium channel gene in *Drosophila*. Science 237:744–749

Sasaki K, Sato M (1987) A single GTP-binding protein regulates K^+-channels coupled with dopamine, histamine and acetylcholine receptors. Nature 325:259–262

Schwarz TL, Tempel BL, Papazian DM, Jan LY (1988) Multiple potassium-channel components are produced by alternative splicing at the Shaker locus in *Drosophila*. Nature 331:137–142

Schwarz W, Passow H (1983) Ca^{2+}-activated K^+ channels in erythrocytes and excitable cells. Annu Rev Physiol 45:359–374

Shuster MJ, Camardo JS, Siegelbaum SA, Kandel ER (1986) Modulation of the 'S' K^+ channel by cAMP-dependent protein phosphorylation in cell-free membrane patches. Prog Brain Res 69:119–132

Siegelbaum SA (1987) The S-current a background potassium current. In: Kaczmarek LK, Levitan IB (eds) Neuromoulation: the biochemical control of neuronal excitability. Oxford University Press, New York

Siegelbaum SA, Camardo JS, Kandel ER (1982) Serotonin and cyclic AMP close single K channels in *Aplysia* sensory neurones. Nature 299:413–417

Sigworth FJ, Sine (1987) Data transformations for improved display and fitting of single-channel dwell time histograms. Biophys J 52:1047–1052

Sims SM, Singer JJ, Walsh JV jr (1985) Cholinergic agonists suppress a potassium current in freshly dissociated smooth muscle cells of the toad. J Physiol (Lond) 367:503–529

Spruce AE, Standen NB, Stanfield PR (1985) Voltage-dependent ATP-sensitive potassium channels of skeletal muscle membranes. Nature 316:736–738

Squire LG, Petersen OH (1987) Modulation of Ca^{2+}- and voltage-activated K^+ channels by internal Mg^{2+} in salivary acinar cells. Biochim Biophys Acta 899:171–175

Stampe P (1985) Proton inactivation of the Ca^{2+}-activated K^+ channel in human red cells. Acta Physiol Scand [Suppl 1] 542:124–162

Stampe P, Vestergaard-Bogind B (1985) The Ca^{2+}-sensitive K^+-conductance of the human red cell membrane is strongly dependent on cellular pH. Biochim Biophys Acta 815:313–321

Stanfield PR (1987) Nucleotides such as ATP may control the activity of ion channels. Trends Neurol Sci 10:335–339

Stanfield PR (1988) Intracellular Mg^{2+} may act as a co-factor in ion channel function. Trends Neurosci 11:475–477

Stanfield PR, Nakajima Y, Yamaguchi K (1985) Substance P raises neuronal excitability by reducing inward rectification. Nature 315:498–501

Storm JF (1987) Action potential repolarization and a fast after-hyperpolarization in rat hippocampal pyramidal cells. J Physiol (Lond) 365:733–759

Tabares L, Lopez-Barneo J, de Miguel C (1985) Calcium- and voltage-activated potassium channels in adrenocortical cell membranes. Biochim Biophys Acta 814:96–102

Takumi T, Ohkubo H, Nakanishi S (1988) Cloning of a membrane protein that induces a slow voltage-gated potassium current. Science 242:1042–1045

Tanabe T, Takeshima H, Mikami A, Flockerzi V, Takahashi H, Kangawa K, Kojima M, Matsuo H, Hirose T, Numa S (1987) Primary structure of the receptor for calcium channel blockers from skeletal muscle. Nature 328:313–318

Taylor PS (1987) Selectivity and patch measurements of A-current channels in *Helix aspersa* neurons. J Physiol (Lond) 388:437–447

Tempel BL, Papazian DM, Schwarz TL, Jan YN, Jan LY (1987) Sequence of a probable potassium channel component encoded at Shaker locus of *Drosophila*. Science 237:770–775

Tempel BL, Jan YN, Jan LY (1988) Cloning of a probable potassium channel gene from mouse brain. Nature 332:837–839

Timpe LC, Schwarz TL, Tempel BL, Papazian DM, Jan YN, Jan LY (1988) Expression of functional potassium channels from Shaker cDNA in *Xenopus* oocytes. Nature 331:143–145

Tokimasa T (1985) Intracellular Ca^{2+} ions inactivate K^+ current in bullfrog sympathetic neurones. Brain Res 337:386–391

Trautmann A, Marty A (1984) Activation of Ca-dependent K channels by carbamoyl choline in rat lacrimal glands. Proc Natl Acad Sci USA 81:611–615

Trube G, Hescheler J (1983) Potassium channels in isolated patches of cardiac cell membrane. Naunyn-Schmiedebergs Arch Pharmacol 322:R64

Trube G, Hescheler J (1984) Inward-rectifying channels in isolated patches of the heart cell membrane: ATP-dependence and comparison with cell-attached patches. Pflügers Arch 401:178–184

Tsuji S, Minota S, Kuba K (1987) Regulation of two ion channels by a common muscarinic receptor-transduction system in a vertebrate neuron. Neurosci Lett 81:139–145

Ubl J, Murer H, Kolb HA (1988) Ion channels activated by osmotic and mechanical stress in membranes of opossum kidney cells. J Membr Biol 104:223–232

Vandenberg CA (1987) Inward rectification of a potassium channel in cardiac ventricular cells depends on internal magnesium ions. Proc Natl Acad Sci USA 84:2560–2564

Vergara C, Latorre R (1983) Kinetics of Ca^{2+}-activated K^+ channels from rabbit muscle incorporated into planar lipid bilayers: evidence for a Ca^{2+} and Ba^{2+} blockade. J Gen Physiol 82:543–568

Vergara C, Moczydlowski E, Latorre R (1984) Conduction, blockade and gating in a Ca^{2+}-activated K^+ channel incorporated into planar lipid bilayers. Biophys J 45:73–76

Volterra A, Siegelbaum SA (1988) Role of two different guanine nucleotide-binding proteins in the antagonistic modulation of the S-type K^+ channel by cAMP and arachidonic acid metabolites in *Aplysia* sensory neurons. Proc Natl Acad Sci USA 85:7810–7814

Wagoner PK, Oxford GS (1987) Cation permeation through the voltage-dependent potassium channel in the squid axon. Characteristics and mechanisms. J Gen Physiol 90:261–290

Watanabe K, Gola M (1987) Forskolin interaction with voltage-dependent K channels in *Helix* is not mediated by cyclic nucleotides. Neurosci Lett 78:211–216

Weiss JN, Lamp ST (1987) Glycolysis preferentially inhibits ATP-sensitive K^+ channels in isolated guinea pig cardiac myocytes. Science 238:67–69

Wong BS, Adler M (1986) Tetraethylammonium blockade of calcium-activated potassium channels in clonal anterior pituitary cells. Pflügers Arch 407:279–289

Wong BS, Lecar H, Adler M (1982) Single calcium-dependent potassium channels clonal anterior pituitary cells. Biophys J 39:313–317

Woodhull AM (1973) Ionic blockade of sodium channels in nerve. J Gen Physiol 61:687–708

Yatani A, Codina J, Brown AM, Birnbaumer L (1987) Direct activation of mammalian atrial muscarinic potassium channels by GTP regulatory protein G_K. Science 235:207–211

Yellen G (1984) Ionic permeation and blockade in Ca^{2+}-activated K^+ channels of bovine chromaffin cells. J Gen Physiol 84:157–186

Yellen G (1987) Permeation in potassium channels: implications for channel structure. Annu Rev Biophys Biophys Chem 16:227–246

Young JA, van Lennep EW (1979) Transport in salivary and salt glands. In: Giebisch G (ed) Transport organs. Springer, Berlin, Heidelberg, New York, pp 563–692 (Membrane transport in biology, vol 4B)

Ypey DL, Clapham DE (1984) Development of a delayed-outward rectifying K^+ conductance in cultured mouse peritoneal macrophages. Proc Natl Acad Sci USA 81:3083−3087

Ypey DL, Ravesloot HP, Buisman HP, Nijweide PJ (1988) Voltage-activated ionic channels and conductances in embryonic chick osteoblast cultures. J Membr Biol 101:141−150

Zhang L, Krnjevic K (1987) Apamin depresses selectively the afterhyperpolarization of cat spinal motoneurons. Neurosci Lett 74:58−62

Rev. Physiol. Biochem. Pharmacol., Vol. 115
© Springer-Verlag 1990

Peptide Toxins and Potassium Channels

FLORIAN DREYER

Contents

Rudolf-Buchheim-Institut für Pharmakologie, Justus-Liebig Universität, Frankfurter Str. 107, D-6300 Gießen, FRG

1 Introduction

After the first detailed description of the delayed outward potassium current in squid axon by Hodgkin and Huxley (1952) it took electrophysiologists more than 20 years to realize that in addition to it several types of K^+ currents can exist in the same cell and that they have a number of functions including modulation of cell excitability and modulation of integrating neuronal function. That as late as 1985 an international symposium on membrane control of cellular activity completely excluded K^+ currents highlights the comparatively late interest of electrophysiologists in these ion currents. Moreover, our knowledge of the structure and function of K^+ channels was very limited compared with what we know about Na^{2+} and Ca^{2+} channels and the acetylcholine receptor. One major reason has been the lack of suitable ligands that act on K^+ channels with high affinity and selectivity such as tetrodotoxin on Na^+ channels, dihydropyridines on one type of Ca^{2+} channels and α-bungarotoxin on nicotinic acetylcholine receptors. This has now profoundly changed mainly due to the development of the patch-clamp recording technique by Neher and Sakmann (Hamill et al. 1981), giving the tool to differentiate between an increasing number of receptor- and/or voltage-operated ion channels. Particularly the K^+ channels have now mushroomed into a large, branching family of at least 11 members (Cook 1988). The dilemma is now that cell membranes, not only excitable ones, are equipped with a variety of K^+ channels differing in their gating properties, which often complicates the interpretation of whole-cell K^+ currents and their modulation by drugs. This difficulty underlines the importance to find both potent and selective ligands for identification and purification of different classes of K^+ channels. The ligands should serve as substrates for radioactive labelling, for raising antibodies and as tools for photoaffinity labelling. Peptides might be particularly useful for such purposes. Description over the past few years of the peptide toxins such as apamin, dendrotoxin and charybdotoxin as blockers of certain types of K^+ channels has made it clear that nature did not forget to equip at least some familiar venomous creatures such as bees, snakes and scorpions with K^+ channel toxins. This aroused the interest of electrophysiologists and biochemists in venom research and venom researchers were sought after. However, to demonstrate the enormous development in and the power of molecular biology and gentechnology, complete amino acid sequences of functional "A-type" and delayed rectifier K^+ channels which have been expressed from *Shaker* cDNA of *Drosophila melanogaster* (Timpe et al. 1988) or of rat brain cDNA (Stühmer et al. 1988) in *Xenopus* oocytes, have been determined without using any specific high affinity label.

I will praise deadly creatures since their venoms are important probes for the study of ion channels in cell membranes. This review deals with those peptide toxins that have been found to be specific for the K^+ channel family. Two excellent reviews dealing with the same subject appeared when this article was in preparation (Moczydlowski et al. 1988; Castle et al. 1989).

2 Diversity of Potassium Channels

2.1 Physiological Properties

With modern electrophysiological techniques an incredible and nearly terrifying number of different types of K^+ channels were identified in excitable, but surprisingly also in non-excitable cell membranes. All cell membranes so far studied exhibit at least two, often multiple distinct types of K^+ channels, but the ratio of their frequency varies widely. This probably accounts for the rather diverse macroscopic K^+ currents found in various types of cells. K^+ channels and their macroscopic currents have been classified according to the inputs required for their gating. Figure 1 shows a simplified classification scheme with some generalized characteristic features.

The "classical" *K^+ currents* are *activated by membrane-potential changes*. They can be subdivided into 3 types: (1) The delayed (outward) rectifying current I_K activated by depolarization of the membrane. (2) An inward (or anomalous) rectifying current I_{IR} which is produced by a rapid conductance increase when the membrane potential is hyperpolarized beyond the potassium equilibrium potential E_K. The outward current is inhibited due to ion channel block by intracellular Mg^{2+}. (3) A fast transient outward current I_A (the A-current) that increases upon membrane depolarization and that is already activated close to the resting potentials (activation range -70 to $-50\,mV$).

At least in the frog node of Ranvier the delayed rectifying K^+ current consists of three different components which can be distinguished on the basis of their kinetic behaviour and voltage dependence (Dubois 1981). This dis-

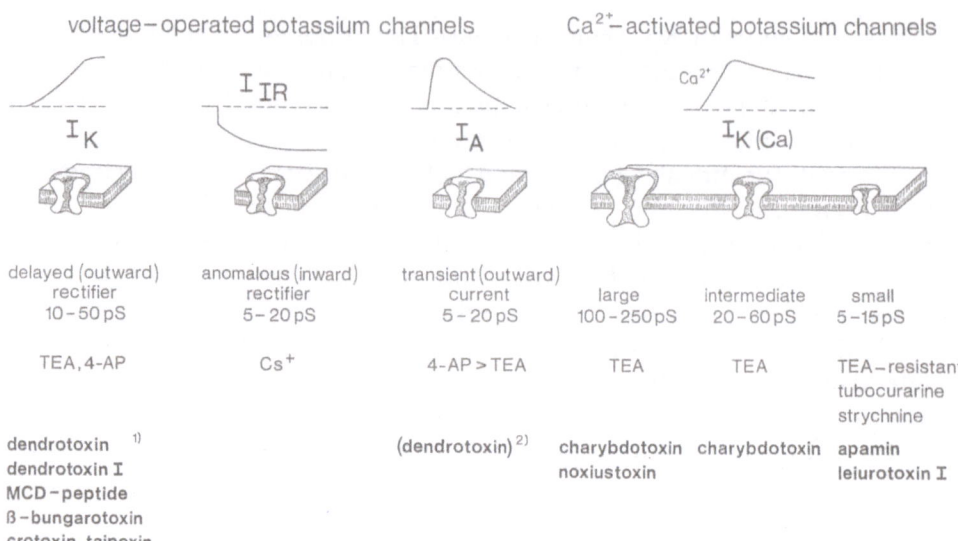

Fig. 1. Classification of the main types of voltage-operated and Ca^{2+}-activated K^+ channels according to their biophysical and pharmacological properties. Top row shows typical K^+ currents eliciated by appropriate changes of membrane potential and/or intracellular Ca^{2+} concentration. Below the values of unitary channel conductance (which should be regarded as a guideline only) some common blocking substances and those peptide toxins that have found to be specific for K^+ channels are listed. (1) In the case of the outward rectifying K^+ channel the toxins listed inhibit only one subtype which most probably exist only in neuronal membrane. (2) In contrast, an inhibition of an I_A-current by the toxin has been shown so far only in hippocampal neurones using high toxin concentrations. Furthermore, a transient, Ca^{2+}-dependent K^+ current present in magnocellular neurosecretory neurones of rat supraoptic nucleus, is dendrotoxin-sensitive in nanomolar concentrations. Tetraethylammonium (TEA), 4-aminopyridine (4-AP), cesium (Cs)

tinction was confirmed by their different sensitivities to toxins (Benoit and Dubois 1986) and pharmacological agents (Dubois 1982). Also the properties of "A-currents" are different in various cell types. In *Drosophila* two distinct K^+ channels underlying the "A-current" in cultured myotubes and neurones have been characterized differing in their conductance, voltage dependence and gating kinetics (Solc et al. 1987).

The next class are the *ligand-operated K^+ channels* which can further be subdivided: (1) Ca^{2+}-activated K^+ channels which open following increases in intracellular Ca^{2+}-concentration. (2) Receptor-controlled K^+ channels directly activated or inactivated via associated guanine nucleotide binding (G) proteins or, physically distinct from the receptor, via second messengers. (3) K^+ channels gated by intracellular binding of the nucleotide ATP. Only Ca^{2+}-activated K^+ channels have been incorporated in Fig. 1 because for the latter two groups no peptide toxin has so far been shown to act on these channels although this cannot be excluded.

The existence of at least three different types of Ca^{2+}-activated K^+ channels is now well established. Those channels exhibiting a high unitary conductance of about $100-250$ pS are often called "Big" or "Maxi" K^+ channels. "Big" and "Intermediate" channels require both Ca^{2+} $(0.1-10\,\mu M$ intracellularly) and membrane depolarization to open. The "Small" channels with a conductance of about 15 pS are activated by cytosolic Ca^{2+} only and show a very low voltage dependence. They are, in contrast to all other types of K^+ channels, insensitive to tetraethylammonium (TEA). The single channel conductances given in Fig. 1 should only be regarded as a guideline. A detailed description of the biophysical properties of the different K^+ channels is given in the preceeding chapter by H.-A. Kolb (see p. 51 ff.) and in other recent reviews (Hille 1984; Rudy 1988; Castle et al. 1989).

2.2 Agents Affecting K^+ Channels

K^+ channels are protein molecules that span cell membranes. The ion channels exist in closed and open conformations and when open ions flow through the channels causing single channel currents of about few picoamperes. Different mechanisms have been shown, or suggested, how drugs interact with K^+ channels leading frequently to a decrease in K^+ current. Some of these modes of action are illustrated in Fig. 2 and may be summarized as follows:

1. Binding of drug molecules to open ion channels inhibits the flow of ions as long as the blocking drug is bound, but also prevents the closing of the channels.

Interactions of drugs with K^+ channels

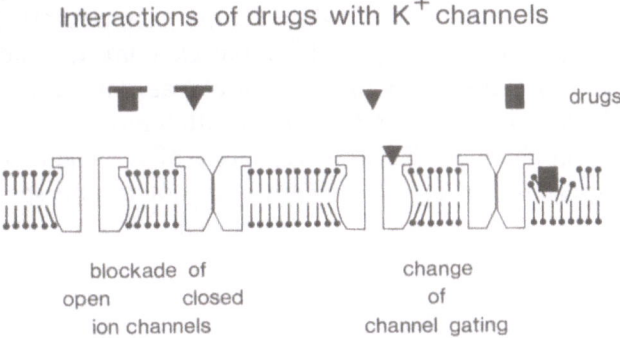

Fig. 2. Possible modes of interaction of drugs with voltage- and/or Ca^{2+}-activated K^+ channels. *Left*: Drugs may block directly open or closed ion channels. *Right*: Change in channel gating that can result in an increase or a decrease of channel open time or probability of channel opening; this may result from drug binding to one or more allosteric sites at subunits of the channel protein complex or from alterations of the lipid environment of the channel

2. Binding of drug molecules to the ion channel in its closed conformation prevents the opening of the channel.
3. Binding of drug molecules to some allosteric sites of the channel protein changes the channel gating. This can lead to a decrease of the channel open time or of the probability of channel opening (K^+ channel blocker), or instead to an increase of the probability of channel opening (K^+ channel opener).
4. Alteration of the lipid environment of the K^+ channel by lipid-soluble compounds may change the gating and conductance of the ion channels ('anaesthetic action').

Drugs which simply block the channels (mechanism 1 and 2) are often called K^+ channel blockers, while the mechanisms 3 and 4 describes drugs that may be named "gate modifiers" of K^+ channels. With receptor-operated K^+ channels, not shown in Fig. 2, further modes of action are the interaction of drug molecules with the recognition site of the receptor itself, with a G-protein, coupling receptor and channel protein, or with a protein kinase coupled to the K^+ channel complex.

Only few substances are available which show a high specificity for K^+ channels. Classical drugs known to have distinct pharmacological effects are TEA and derivatives of aminopyridine. It is generally believed that TEA is usually more effective in blocking non-inactivating K^+ currents, whereas aminopyridines affect preferentially transient, fast-inactivating K^+ currents. However, the specificities are not always clear cut, especially when high concentrations of these drugs are employed. It is therefore impossible to use either of the substances to block a specific type of K^+ channel completely without affecting other K^+ channels. Further, their affinities differ tremendously between different animal species and different types of cells (for reviews, see Stanfield 1983; Hille 1984; Rogawski 1985; Cook 1988; Rudy 1988; Castle et al. 1989). Other compounds such as quinine, local anaesthetics, non-depolarizing muscle relaxants and some ions like Cs^+ and Zn^{2+} also effectively block all or at least some types of K^+ channels, but unfortunately they have additional, often more specific effects on other classes of ion channels. This heterogeneity of K^+ channel blockade underlines the importance of studies to find both potent and specific agents and toxins.

3 Toxins from Bee Venom

Venom of the bee (*Apis mellifera*) is composed of several toxic components which have been purified to homogeneity and extensively characterized by Habermann and coworkers (for reviews, see Habermann 1968, 1972). Besides histamine, hyaluronidase and phospholipase A_2 the venom contains also some peptide toxins like melittin, apamin and the mast cell degranulating (MCD) peptide. Apamin and MCD peptide constitute not more than 2% each of the crude venom. The toxins are relatively small peptides compared with neurotoxins of other venomous animals such as snakes, scorpions and spiders. Apamin (MW 2039) is a basic, single chain peptide consisting of 18 amino acids and MCD peptide (MW 2593) is composed of 22 amino acids. Both toxins contain two disulfide bridges and – although they differ by their amino acid sequence – their general structure is very similar. Studies with apamin congeners have shown that the arginine residues at position 14 and 15 are essential for the biological activity. Both toxins show neurotoxicity upon central application (Habermann 1977), but they differ in their potencies and also qualitatively in the symptoms they cause. This suggests different sites of action that was confirmed by recent electrophysiological data (see below).

3.1 Apamin

Since its purification apamin is known as a preferentially if not exclusively centrally acting neurotoxin (Habermann and Reiz 1965). The characteristic symptoms of apamin poisoning consist of extreme uncoordinated hypermotility and hyperexcitability culminating in generalized convulsions. These symptoms can be evoked by peripheral ($LD_{50} \approx 4$ mg/kg mouse intravenously) and intraventricular application, but in the latter case with an about 3000-fold lower concentration. The central effects can be taken as a test of apamin's biological activity (for detailed description, see Habermann 1972, 1984).

3.1.1 Binding of Apamin

An important step forward in the elucidation of apamin action was the iodination of apamin at its histidine residue (Habermann and Fischer 1979). This allowed autoradiographic localization and quantification of binding sites for ^{125}I-apamin in specific parts of the central nervous system of a variety of species (Habermann and Horvath 1980; Janicki et al. 1984; Mourre et al. 1984, 1986). The very high affinity of apamin to rat, guinea pig and canine brain membranes with a dissociation constant K_D in the range of 20 to 50 pM was in agreement with its neurotoxicity (Habermann and Fischer 1979; Hugues et al. 1982a; Wu et al. 1985). The binding to peripheral organs was much less, however with some exceptions, namely the guinea pig liver (Cook et al. 1983) and colon (Hugues et al. 1982b), the rabbit liver and the bovine adrenal cortex (Habermann and Fischer 1979) and rat smooth muscles and hepatocytes (Romey et al. 1984). High affinity binding of apamin was also found in mouse neuroblastoma cells (Hugues et al. 1982c), cultured rat embryonic neurones (Seagar et al. 1984) and in cultured rat muscle cells (Hugues et al. 1982e). Potassium ions served as pronounced and specific promoters of binding (Habermann and Fischer 1979). This observation indicated a functional interaction between the peptide and the cation.

Purification and Characterization of Apamin-Binding Protein

The high-affinity binding of apamin to brain cell membranes is inviting to purify the apamin acceptor, most probably linked to a Ca^{2+}-activated K^+ channel, and to characterize its structure. Using the radiation inactivation technique (target size analysis) Schmid-Antomarchi et al. (1984) reported a molecular mass of about 250000, quite different to the value of about 86000 Da found by Seagar et al. (1986). Affinity labelling of the apamin receptor by crosslinking the toxin with its binding site on the K^+ channel (Hugues et al. 1982d; Schmid-Antomarchi et al. 1984) or by photoaffinity labelling technique (Seagar et al. 1985, 1987a; Marqueze et al. 1987) gave divergent results with polypeptides of different molecular mass which are considered to be subunits of a Ca^{2+}-activated K^+ channel. Also solubilization of the apamin receptor has been performed (Schmid-Antomarchi et al. 1984; Seagar et al. 1987b), but due to its very low density in membrane preparations so far the binding protein could not be purified to homogeneity.

3.1.2 Apamin Blocks One Type of Ca^{2+}-Activated K^+ Channels

Although the neurotoxicity of apamin is pronounced, initial evidence of its putative molecular mechanism of action came from studies on intestinal smooth muscles and isolated hepatocytes of the guinea pig. Apamin was the most selective agent acting in nanomolar concentrations that inhibited α-adrenoreceptor agonist and ATP-induced hyperpolarization in smooth muscle cells (Vladimirova and Shuba 1978; Banks et al. 1979; Maas and Den Hertog 1979; Shuba and Vladimirova 1980) as well as the K^+ loss from hepatocytes (Banks et al. 1979; Burgess et al. 1981). The K^+ loss evoked by the Ca^{2+} ionophore A 23187 was also abolished by apamin whilst the receptor-mediated ^{45}Ca efflux was unaffected by the toxin. Therefore, Jenkinson and co-workers suggested that apamin directly blocks Ca^{2+}-dependent K^+ channels involved in drug action (Banks et al. 1979; Burgess et al. 1981). Their additional finding that apamin did not, however, block Ca^{2+}-dependent K^+ permeability of red blood cells gave initial evidence that at least two different types of Ca^{2+}-activated K^+ channels may exist. That apamin blocks certain K^+ channels was also concluded from studies on adrenaline or ATP-induced $^{42}K^+$ efflux in smooth muscle cells of guinea pig taenia caeci (Maas et al. 1980; Den Hertog 1981).

The first electrophysiological evidence for an action of apamin on neurones was provided by Hugues et al. (1982c) who studied action potential and membrane currents of differentiated neuroblastoma cells (clone N1E 115). In high extracellular Ca^{2+} concentration about 20% of the cells exhibited action potentials followed by prolonged, TEA-insensitive after-hyperpolarizations that were completely blocked by 100 nM apamin. The underlying, TEA-insensitive, Ca^{2+}-dependent K^+ current was suppressed by the toxin with an IC_{50} value of about 10 nM, whereas the fast Na^+ current and the TEA-sensitive delayed outward K^+ current were unaltered. The apamin-sensitive K^+ current which seems to play some role in the control of repetitive discharge patterns of action potentials is easily overlooked when using physiological Ca^{2+} concentration. This is consistent with the finding that apamin binding experiments at the same cell preparation revealed a low binding capacity of about 12 fmol/mg protein, a value about 1/5th of that found for the tetrodotoxin-sensitive Na^+ current (Hugues et al. 1982c). The dissociation constant of the apamin-receptor complex was determined to be about 20 pM. Such a large discrepancy between the effective concentrations of apamin found in physiological measurements and in binding experiments was also reported for the guinea pig colon (Hugues et al. 1982b). The discrepancy seems typical for most toxin studies and may arise because binding studies are performed mostly at unphysiological conditions like using cell homogenates and solutions of low ionic strength. Avoiding this Seagar et al. (1984) at rat embryonic neurones and Cook and Haylett (1985) at guinea pig hepatocytes demonstrated a fairly good agreement within a factor of 3 between the pharmacological concentration-response curve of native apamin and the competition between ^{125}I-apamin and native apamin in binding experiments.

Apamin has further been shown to block Ca^{2+}-dependent K^+ currents in cultured rat muscle cells (Hugues et al. 1982e) and the Ca^{2+} ionophore-induced ^{86}Rb release from cultured rat embryonic neurones (Seagar et al. 1984). The toxin, however, did not block Ca^{2+}-activated K^+ currents in erythrocytes (Burgess et al. 1981), in molluscan neurones (*Aplysia, Helix*) (Hermann and Hartung 1983) and in pancreatic B-cells (Lebrun et al. 1983). Further studies at rat myotubes (Romey and Lazdunski 1984; Romey et al. 1984) and bullfrog sympathetic ganglia (Pennefather et al. 1985) revealed then the existence of at least two different kinds of Ca^{2+}-dependent K^+ currents. One type, responsible for the long-lasting after-hyperpolarization, is blocked by apamin, but refractory to TEA, whereas the other one is blocked, just vice versa, by TEA, but not by apamin.

3.1.3 Properties of the Apamin-Sensitive K^+ Channels

That the different electrophysiological and pharmacological properties of Ca^{2+}-activated K^+ currents are really due to quite distinct types of K^+ channels is strongly supported by two findings. Firstly, a scorpion peptide toxin, charybdotoxin, specifically blocks a Ca^{2+}-activated K^+ channel that is dominantly characterized by its *large* single channel conductance of 100–250 pS and its strong voltage dependence (Miller et al. 1985). Secondly, Blatz and Magleby (1986) identified the target for apamin in cultured rat skeletal muscle as a Ca^{2+}-activated, TEA-insensitive K^+ channel of *small* conductance (10–14 pS). This type of K^+ channel is further characterized by its weak voltage sensitivity and a ten times higher Ca^{2+} sensitivity at negative membrane potentials compared to the Ca^{2+}-activated K^+ channel of large conductance. These properties of the apamin-sensitive K^+ channel are best suited to generate long-lasting after-hyperpolarization that follows the action potential in many types of cells.

Studies on hepatocytes revealed that apamin-sensitive K^+ channels are also blocked by other agents including quinine and non-depolarizing muscle relaxants (Jenkinson et al. 1983; Cook and Haylett 1985). Tubocurarine, pancuronium and atracurium are the most potent ones, acting in micromolar concentrations. This very characteristic feature correlates well with the ability of these neuromuscular blockers to inhibit the binding of labelled apamin. The explanation is that these neuromuscular blockers have two charged nitrogens the distance between which is similar to that between the two charged arginines in the peptide chain of apamin.

Apamin-sensitive, TEA-insensitive K^+ channels have been described in a variety of other cells, both excitable and inexcitable, yet it is not known if all of them have the same biophysical properties as those studied in detail in rat myotubes. Nevertheless, it is presently almost a dogma that apamin is highly selective for the Ca^{2+}-activated, TEA-resistant, d-tubocurarine-sensitive, nearly voltage-independent K^+ channel of small conductance. So far it is not clear whether the effect of apamin results from a change of K^+ channel gating properties (reduction of the open time, decreased probability of channel opening or both) or from a block of the channel conductance from outside. So far, only two reports deal with the question if apamin can also block the Ca^{2+}-activated K^+ channels from the cytoplasmic side. Zhang and Krnjevic (1987) reported that extra-, but not intracellular injection of the toxin by ionophoresis, abolished selectively the after-hyperpolarizations of cat spinal motoneurones in situ. However, Szente et al. (1988) observed that pressure injection of apamin into cat neocortical neurones in situ led to a selective block of an outward K^+ current that mediated a slow after-hyperpolarization. To solve this discrepancy patch-clamp experiments on inside-out membrane patches must be waited for. Furthermore, to date effects of apamin on other types of ion channels cannot be excluded.

3.1.4 Expression of Apamin-Sensitive K^+ Channels

The apamin-sensitive Ca^{2+}-dependent K^+ channel has been best characterized in cultured rat skeletal muscle cells (Hugues et al. 1982e; Romey and Lazdunski 1984; Blatz and Magleby 1986). The expression of this type of K^+ channel in muscle cells depends on the stage of their innervation (Schmid-Antomarchi et al. 1985). When rat myotubes were cocultured with nerve cells from the rat spinal cord, the action potentials were no longer followed by long-lasting after-hyperpolarizations. Furthermore, while action potentials of innervated rat EDL muscle fibres were not followed by after-hyperpolarizations, this was regularly observed in muscle fibers 8–10 days after denervation. In keeping with these after-hyperpolarizations, apamin was effective in nanomolar concentrations. This was also in line with binding experiments using ^{125}I-labelled apamin (Schmid-Antomarchi et al. 1985). An expression of apamin-binding sites could be demonstrated in rat leg muscles starting after 2 days of transection of the sciatic nerve. Schmid-Antomarchi et al. (1985) followed also the binding capacity for ^{125}I-apamin in rat skeletal muscles during *in utero* and postnatal life. Beginning at 15 days of development the apamin binding capacity decreased as muscle maturation proceeded and completely disappeared around 5 days of the postnatal life. The conclusion was that the apamin-sensitive K^+ channels are fully expressed in denervated mammalian muscle cells, but completely absent in innervated ones.

The myotonic muscular dystrophy, called Steinert disease, is characterized clinically by muscle stiffness and difficulty in muscle relaxation after voluntary contractions. The pathologic muscle fibers tend to fire repetitive action potentials in response to direct electrical and mechanical stimulation. The use of radiolabelled apamin enabled Renaud et al. (1986) to demonstrate that diseased human muscle fibres although still innervated, have expressed apamin binding sites which are completely absent in normal human muscle. Assuming that the apamin binding sites in myotonic dystrophic muscle fibres correspond to apamin-sensitive K^+ channels these ion channels may contribute to the abnormal behaviour to generate repetitive bursts of activity. Which factor, however, controls the expression of this type of Ca^{2+}-activated K^+ channel is still unknown.

Seagar et al. (1984) followed the *in vitro* ontogenesis of apamin receptors in embryonic neurones dissected from brains of 16-day-old rat embryos. They found that apamin binding sites were present at the earliest stage of development and that the number did not change during 8 days in culture. In the same period, however, the binding of the toxin II of the scorpion *Androctonus australis* Hector, specific for Na^+ channels, increased by a factor of 10, suggesting that the ontogenesis of Ca^{2+}-dependent K^+ channels and of voltage-dependent Na^+ channels are independent processes.

Different time courses of ion channel development have also been reported by Mourre et al. (1987). In rat brain these authors followed the postnatal development of (1) the voltage-dependent Na^+ channel, (2) one type of Ca^{2+} channels and (3) one type of Ca^{2+}-activated K^+ channels by autoradiography and by binding studies using tetrodotoxin, verapamil and apamin as specific markers, respectively. The general observation was that the maximal binding capacity of verapamil increased by a factor of 20 from day 0 to the postnatal day 30 while the apamin-sensitive K^+ channels were present in high density without any major change during the same period. Also the affinity of the Ca^{2+}-activated K^+ channel for ^{125}I-apamin remained constant during postnatal ontogenesis while tetrodotoxin binding sites not only increased in density but also showed a decrease in affinity for tetrodotoxin.

3.1.5 Endogenous Apamin-Like Factor

Apparently stimulated by the history of opioid or benzodiazepine receptors and the existence of high-affinity binding sites for apamin, Fosset et al. (1984) looked for an endogenous substance that may use the apamin receptors to modulate cell excitability in neurones. Indeed, they presented evidences for an apamin-like peptide existing in pig brain. The isolated peptide had the same immunological and pharmacological properties as apamin itself. However, the amount that could be purified from one pig brain had an activity that was equivalent to apamin as low as 1.5 ng only. It is tentative to suggest, but still unproven, that this apamin-like factor does play a physiological role by modulating apamin-sensitive Ca^{2+}-activated K^+ channels.

3.1.6 Apamin as a Pharmacological Probe

The many studies in which apamin has been used as a specific probe to distinguish Ca^{2+}-activated K^+ currents are listed in Table 1. In some experiments the dissection was combined with charybdotoxin (see Sect. 4.2.1 and Table 2). Since the finding that apamin blocks responses caused by enteric inhibitory nerves or by noradrenaline and ATP in gut smooth muscle cells, the toxin has been used as a pharmacological tool in a large number of studies on the neural control of the gastrointestinal tract (see for example Kitabgy and Vincent 1981; Goedert et al. 1984; Costa et al. 1986).

3.2 Mast Cell Degranulating Peptide

The results of various studies with mast cell degranulating (MCD) peptide can be summarized by the statement that its action on K^+ channels is most probably identical with the action of the dendrotoxins (see Sect. 5.1; Table 3).

Table 1. Blocking effects of apamin on potassium efflux, currents and channels in various cell preparations

Response which is reduced or abolished by apamin	Cell type	Apamin concentration	References
Smooth Muscles			
Adrenaline/ATP-induced ^{42}K efflux	Taenia caeci of guinea pig	100 nM	Maas et al. 1980; Den Hertog 1981
Neurotensin-induced relaxation	Colon of guinea pig	IC_{50} = 7 nM	Hugues et al. 1982b
^{86}Rb efflux induced by bradykinin, amidephrine	Taenia caeci of guinea pig	<10 nM	Gater et al. 1985
Noradrenaline-induced ^{86}Rb efflux	Taenia caeci of guinea pig	100 nM	Weir and Weston 1986
Adrenaline-induced relaxation	Taenia coli of guinea pig	EC_{50} = 1.2 nM	Chicchi et al. 1988
Skeletal muscles			
Ca^{2}-dependent K$^+$ current	Cultured skeletal muscle of rat	≈10 nM	Hugues et al. 1982e; Romey and Lazdunski 1984; Schmid-Antomarchi et al. 1985
Slow outward K$^+$ current	Skeletal muscle fibres of frog	≈50 nM	Cognard et al. 1984; Traoré et al. 1986
After-hyperpolarization that follows action potential	Denervated EDL muscle of rat	≈10 nM	Schmid-Antomarchi et al. 1985
Ca^{2+}-activated K$^+$ channel of *small* conductance	Cultured skeletal muscle of rat	<1 µM	Blatz and Magleby 1986
Neurones			
After-hyperpolarization and underlying Ca^{2+}-dependent K$^+$ current	Differentiated neuroblastoma cells (clone N1E 115)	IC_{50} ≈10 nM	Hugues et al. 1982c
^{86}Rb release induced by Ca-ionophore	Cultured embryonic neurones of rat	EC_{50} ≈300 pM	Seagar et al. 1984
After-hyperpolarization	Superior cervical ganglia of rat	100 nM	Galvan and Behrends 1985
Ca^{2+}-dependent currents underlying after-hyperpolarization	Sympathetic ganglion cells of bullfrog	≈25 nM	Pennefather et al. 1985; Goh and Pennefather 1987
Ca^{2+}-activated K$^+$ current	Sympathetic ganglion cells of bullfrog	10 nM	Tanaka et al. 1986
Ca^{2+}-dependent after-hyperpolarization	Sympathetic neurones of rat	40 nM	Kawai and Watanabe 1986
Ca^{2+}-activated K$^+$ current	Anterior pituitary cell line of rat	20–100 nM	Ritchie 1987
After-hyperpolarization	Spinal motoneurones of cat	ionophoretic application	Zhang and Krnjević 1987
After-hyperpolarization	Neurosecretory neurones of rat hypo-thalamic supraoptic nucleus	IC_{50} = 1.3 nM	Bourque and Brown 1987
After-hyperpolarization and slow outward K$^+$ current	Neurones of the motor cortex of cat	intracellular injection	Szente et al. 1988

Table 1 (continued)

Response which is reduced or abolished by apamin	Cell type	Apamin concentration	References
	Liver cells		
K^+ efflux induced by noradrenaline, ATP, A 23187	Hepatocytes of guinea pig	≈ 10 nM	Banks et al. 1979; Burgess et al. 1981
K^+ efflux caused by amidephrine, angiotensin II, A 23187	Hepatocytes of guinea pig	$IC_{50} = 1$ nM	Jenkinson et al. 1983; Cook and Haylett 1985
Hyperpolarization caused by noradrenaline	Hepatocytes of guinea pig and rabbit	50 nM	Field and Jenkinson 1987
	Other types of cells		
^{86}Rb efflux caused by noradrenaline, A 23187	Brown adipocytes of hamster	$EC_{50} = 2$ nM	Nanberg et al. 1985
Electrical burst activity induced by glucose	Pancreatic β-cells of OB/OB mice	$40 - 400$ nM	Rosario 1985
	No effect on pancreatic islets of albino mice		Lebrun et al. 1983
$^{42}K^+$ efflux induced by amino acid or sugar	Enterocytes from rabbit jejunum	500 nM	Brown and Sepúlveda 1985
^{86}Rb efflux induced by Ca-ionophore A 23187	Undifferentiated rat pheochromocytoma (PC12) cells	$IC_{50} = 240$ pM	Schmid-Antomarchi et al. 1986

In contrast to apamin, MCD peptide is hardly neurotoxic upon intravenous administration in mice, but following intraventricular injections it causes epileptiform seizures leading to death. The LD_{50} ($\approx 10\,\mu g/kg$ mouse) is approximately 10 times higher than that of apamin (Habermann 1977). Quite distinct from its action on K^+ channels is the degranulation of mast cells (hence its name) and the histamine release (Breithaupt and Habermann 1968) that can be taken as a test of the biological activity of MCD peptide.

3.2.1 Binding of Mast Cell Degranulating Peptide to Brain Membranes

Toxin-binding sites of high affinity ($K_D = 0.15$ nM) that could be copurified with synaptosomal membranes, are distributed throughout the rat brain (Taylor et al. 1984). Quantitative autoradiography of brain slices from rats reveals that the toxin-binding sites are localized mainly in the stratum radiatum of Ammon's horn, neocortex, molecular layer of the cerebellum, colliculi and pons (Bidard et al. 1987a; Mourre et al. 1988). Chemical modification of the toxin molecule always resulted in a complete loss of its binding properties (Taylor et al. 1984).

While distinct binding sites for MCD peptide and apamin exist in synaptosomal membranes of rat brain (Taylor et al. 1984), common acceptor sites have been reported for MCD peptide and the snake venom toxins dendrotoxin and dendrotoxin I (Bidard et al. 1987b; Stansfeld et al. 1987; Rehm et al. 1988) and β-bungarotoxin (Schmidt et al. 1988). However, while Bidard et al. (1987b) found that MCD peptide and dendrotoxin I inhibit binding of ^{125}I-labelled MCD peptide with K_{50}-values of 0.25 and 0.16 nM, respectively, Stansfeld et al. (1987) found a much weaker affinity of MCD peptide for these acceptors. Binding studies mostly give about 100 fold higher binding affinities compared to effective concentrations necessary in electrophysiological experiments. If binding experiments, however, are performed at intact brain slices they reveal about 10 times higher toxin concentrations to obtain half-maximal inhibition, closing the large gap between binding and physiological experiments (Bidard et al. 1987b).

Binding studies with nerve fibre membranes of *Xenopus* reveal high-affinity binding sites for ^{125}I-dendrotoxin which could be displaced from its sites completely by unlabelled MCD peptide (Bräu et al. 1990). Competition experiments performed in Ringer solution allowed to calculate an apparent equilibrium dissociation constant K_I of 1.8 nM for MCD peptide. Similar results were obtained with nerve fibre membranes of rat (H. Repp and F. Dreyer, unpublished data).

By cross-linking experiments with MCD peptide and dendrotoxin I the toxin-binding sites have been identified and solubilized with almost identical molecular weights (about 77000) and properties (Rehm et al. 1988). Studying then binding of iodinated dendrotoxin I and MCD peptide to their solubilized acceptor proteins, Rehm et al. (1988) presented evidences that the two different toxins act on the same protein complex that is assumed to be part of a K^+ channel.

3.2.2 Mast Cell Degranulating Peptide Blocks a Subtype of the Delayed Outward Rectifier Potassium Current

The first electrophysiological study with MCD peptide was conducted by Stansfeld et al. (1987) on peripheral neurones using A cells of rat nodose ganglia. Their finding that MCD peptide has a dendrotoxin-like action on the same type of K^+ currents was surprising because the toxin does not show any structure homology to dendrotoxin. Like the dendrotoxins (Sect. 5.1) MCD peptide induces repetitive generation of action potentials in response to depolarizing currents, which in the control are subthreshold for spike generation. Under voltage-clamp conditions it inhibits an almost non-inactivating outward rectifier K^+ current with an IC_{50}-value of 37 nM that is 17 fold higher than for dendrotoxin. These differences in potency are close to those found for the binding affinities (Stansfeld et al. 1987). In contrast, the transient, rapidly inactivating A-current, present in C cells of nodose ganglia as well as in superior cervical ganglia, is not affected by either toxins (Stansfeld et al. 1987).

Mammalian motor nerve terminals exhibit at least three different types of K^+ currents. Here MCD peptide (3–300 nM), like the snake toxins dendrotoxin and β-bungarotoxin (see Table 3), affects only one type, i.e. the slowly activating outward K^+ current, whereas the fast transient and the Ca^{2+}-activated K^+ currents remain unaffected (J. Beise and F. Dreyer, unpublished data). In frog myelinated nerve fibres MCD peptide inhibits the dendrotoxin-sensitive component I_{Kf1} of the delayed outward K^+ current with an IC_{50}-value of 33 nM (Bräu et al. 1990). Recently MCD peptide-sensitive K^+ channels with delayed rectifier properties have been expressed from rat brain cDNA in *Xenopus* oocytes (Stühmer et al. 1988). The K^+ channels are blocked by the same MCD peptide concentrations as found for the K^+ current in the frog node of Ranvier (Table 3; Sect. 5.1.3).

So far all biochemical and electrophysiological data indicate that MCD peptide and dendrotoxins although structurally unrelated peptides, act on the same subtype of the delayed outward K^+ current in a variety of tissues. They have the same efficacy, but a quite different potency. Therefore, the main features of the MCD peptide-sensitive K^+ channel are identical with those described in detail for dendrotoxin (see Sect. 5.1.6). The mast cell degranulating activity and the blockade of K^+ channels are quite distinct modes of action of MCD peptide in accordance with the finding that dendrotoxin is unable to degranulate mast cells or to serve as an inhibitor of MCD peptide-induced histamine release.

3.2.3 Mast Cell Degranulating Peptide as a Pharmacological Probe

As MCD peptide produces arousal at low doses and convulsions at higher ones and binds with high density to high-affinity acceptor sites, particularly in hippocampus, Cherubini et al. (1987, 1988) were interested in the influence of the toxin on neuronal activity in the CA1 and CA3 region of rat hippocampal slices. Indeed, in the CA1 region the toxin causes long-term potentiation of synaptic transmission which is undistinguishable from that caused by trains of high-frequency nerve stimulation. In the CA3 region the toxin produces long-lasting spontaneous and evoked bursts of action potentials of epileptiform pattern.

In search for an endogenous ligand for the high-affinity binding sites of MCD peptide Cherubini et al. (1987) isolated from pig brain a peptide which was pharmacologically and immunologically similar to MCD peptide. Although suggesting some physiological importance the amount that could be purified per one brain was as low as 0.7 ng.

By quantitative autoradiography and biochemical analysis Mourre et al. (1988) followed the expression of MCD peptide acceptors in rat brain during postnatal life. The density of toxin-binding sites is low during the perinatal period, but increased rapidly by postnatal day 10 accompanied by a decrease of binding affinity. At postnatal day 30 the adult state of toxin distribution in the rat brain is reached. This development is correlated with an increased neurotoxicity of MCD peptide.

4 Toxins of Scorpion Venoms

Scorpion venoms contain multiple toxins as do those of other venomous animals such as snakes, spiders and bees. As early as 1968 Koppenhöfer and Schmidt presented initial evidence showing that the crude venom of the mideastern yellow scorpion *Leiurus quinquestriatus* affects both Na^+ and K^+ permeabilities in the frog node of Ranvier. Similar effects were observed with the crude venom of the scorpion *Buthus tamulus* on voltage-clamped squid axon membrane (Narahashi et al. 1972). The first report that a purified scorpion toxin depresses selectively a voltage-dependent K^+ current in the squid giant axon was made by Carbone et al. (1982).

4.1 Noxiustoxin from the Mexican Scorpion *Centruroides noxius*

Noxiustoxin, a minor component (2% of the total venom protein) of the venom of the Mexican scorpion *Centruroides noxius*, is a short-chain polypeptide with a molecular mass of about 4000 consisting of 39 amino acids (Possani et al. 1982). Its action on K^+ current of squid giant axon was quite selective (no significant effect on Na^+ current), its potency, however, is low (K_D of about 400 nM) and its action is reversible (Carbone et al. 1982). At concentrations below 1.5 µM noxiustoxin blocked the K^+ current independent of the membrane potential and with little effect on the kinetic behaviour (Carbone et al. 1987). At higher concentrations the block by the toxin became more complex depending now on membrane potential and the frequency of pulsing. The conclusion was that noxiustoxin interacts directly with the open K^+ channel.

Single channel recording provided evidence that noxiustoxin, like charybdotoxin (Sect. 4.2.1), affects also Ca^{2+}-activated K^+ channels from the transverse-tubules of skeletal muscle, yet with a 200 times lower potency (Valdivia et al. 1988). The apparent K_D of 450 nM for noxiustoxin was the same as reported for its effect on the delayed rectifying current of the squid axon. While charybdotoxin, due to its much higher potency, produced long-term blockade of K^+ channels, noxiustoxin reduced the open probability by causing brief closed states. The application of both toxins revealed that they can obviously interact simultaneously with the same K^+ channel (Valdivia et al. 1988).

In the central nervous system the potency of noxiustoxin may be much higher for as yet unspecified K^+ channels. Sitges et al. (1986) reported that noxiustoxin blocked the efflux of ^{86}Rb from mouse brain synaptosomes and also increased the GABA-release, probably due to the blockade of a K^+ permeability. The EC_{50}-value was about 3 nM which is much lower than that found for the squid axon ($EC_{50} = 400$ nM).

4.2 Toxins from the Scorpion *Leiurus quinquestriatus*

4.2.1 Charybdotoxin

Charybdotoxin has gained wide acceptance as a specific blocker of a Ca^{2+}-activated K^+ channel (Miller et al. 1985). The toxin is a minor component constituting about 0.1% of the total venom protein of the mideastern scorpion *Leiurus quinquestriatus* var. *hebraeus*. After some discrepancies in molecular mass determination it has now been settled that charybdotoxin is a basic single-chain polypeptide of 4353 Da consisting of 37 amino acids (Gimenez-Gallego et al. 1988; Valdivia et al. 1988). Its primary sequence shows a high homology to noxiustoxin of the scorpion *Centruroides noxius* (Sect. 4.1) and to a number of other neurotoxins of diverse origin including snakes and marine worms. Charybdotoxin may represent a member of a wider family of proteins that modify ion channel activity (Gimenez-Gallego et al. 1988). On the other hand, little structure homology was found with other well known K^+ channel blockers such as apamin that effects TEA-insensitive, Ca^{2+}-activated K^+ channels of small unitary conductance, or dendrotoxin and MCD-peptide that block one component of the delayed outward current. Nevertheless we should always keep in mind that structure homology between toxins is not necessarily correlated to their physiological effects.

Charybdotoxin was introduced as a potent blocker of the high-conductance (≈ 200 pS) Ca^{2+}-activated K^+ channel in transverse tubule membrane of the rat skeletal muscle (Miller et al. 1985). Kinetic analysis of single channel recordings revealed a mode of action in which one charybdotoxin molecule occludes the K^+ channel by binding to its external mouth similar to the mechanism by which tetrodotoxin and saxitoxin block Na^+ channels (Miller et al. 1985; Smith et al. 1986). The idea of physical occlusion of the ion channel is further supported by competition experiments between charybdotoxin and TEA. The results demonstrated that when TEA blocks the K^+ channel from the outside, charybdotoxin cannot attain its binding site (Miller 1988). The toxin molecules interact with the K^+ channels in a simple bimolecular reaction regardless whether the channels are in their open or closed conformation (Anderson et al. 1988, MacKinnon and Miller 1988). The association rate strongly depends on the ionic strength of the external medium while the dissociation of the toxin molecule from its binding site at the channel is enhanced by depolarization. The appearance of long-term non-conducting states is consistent with the high affinity of the toxin for its K^+ channels.

High-conductance Ca^{2+}-activated K^+ channels have been described in a variety of cells of neuronal and non-neuronal origin (see preceding chapter by A. Kolb). This type of K^+ channels, found in bovine aortic smooth muscle cells or GH_3 pituitary cells, were used by Gimenez-Gallego et al. (1988) to follow the biological activity of charybdotoxin during its purification. The effective EC_{50} of purified toxin was 2 nM for both types of cells. In neurones of the marine mollusc *Aplysia californica* Ca^{2+}-activated K^+ channels of intermediate unitary conductance (≈ 35 pS) are also sensitive to charybdotoxin (Hermann and Erxleben 1987). Like in the studies with GH_3 pituitary cells the toxin affected neither the transient nor the delayed outward K^+ currents nor the Na^+ and Ca^{2+} inward currents.

However, charybdotoxin inhibits also Ca^{2+}-insensitive K^+ channels. Two types of voltage-activated, Ca^{2+}-insensitive K^+ channels (unitary conductance 18 pS) from murine T lymphocytes were found to be completely blocked by 5 nM charybdotoxin (Lewis and Cahalan 1988) as well as the *Drosophila Shaker* K^+ channels underlying A-current, expressed in *Xenopus* oocytes (MacKinnon et al. 1988). Recently Schweitz et al. (1989) reported that in rat dorsal root ganglion cells the toxin also blocks voltage-activated K^+ channels that are known to be the target of other peptide neurotoxins such as dendrotoxin and MCD peptide. Also in frog nodal membrane charybdotoxin inhibits the component f1 of the delayed outward K^+ current and displaces ^{125}I-labelled dendrotoxin from its binding sites in nanomolar concentrations (Bräu et al. 1990). This suggest structural similarities between K^+ channels of quite different properties. However, dendrotoxin did not inhibit charybdotoxin-sensitive, Ca^{2+}-activated K^+ channels in smooth muscle cells of GH_3 pituitary cells (see Sect. 5.1.7). So far charybdotoxin maintains only its exclusive action on K^+-specific channels.

The advantage of a probe for Ca^{2+}-activated K^+ channels is best demonstrated by several studies in rat hippocampal neurones (Lancaster and Nicoll 1987; Storm 1987; Alger and Williamson 1988) and in bullfrog ganglion neurones (Goh and Pennefather 1987) dealing with the analysis of action potential-repolarization and afterhyperpolarization currents. The use of charybdotoxin and apamin made it possible to differentiate between two distinct types of Ca^{2+}-activated K^+ currents involved in the repolarization of action potentials.

4.2.2 Leiurotoxin I

Venom of *Leiurus quinquestriatus* inhibits ^{86}Rb uptake through apamin-insensitive channels of human erythrocytes, Ehrlich cells and rat thymocytes and through the apamin-sensitive channels of guinea pig hepatocytes (Abia et al. 1986). Castle and Strong (1986) used apamin as a tool to identify and to partially purify a protein component in the venom of the scorpion *Leiurus quinquestriatus hebraeus* which inhibited apamin binding to and blocked angiotensin II-induced K^+ efflux from guinea pig hepatocytes. This component has been purified to homogeneity and named leiurotoxin I (Chicchi et al. 1988). The toxin is a very basic, single-chain peptide consisting of 31 amino acids with a molecular mass of 3.4 kDa and represents less than 0.02% of the venom protein. It inhibits ^{125}I-apamin binding to rat brain synaptosomal membranes ($K_i = 75$ pM) and blocks the adrenaline-induced relaxation of guinea pig taenia coli ($EC_{50} = 6.5$ nM) (Chicchi et al. 1988).

Although leiurotoxin I and apamin are structurally unrelated peptides they both seem to act on the same Ca^{2+}-activated K^+ channel of *small* conductance, but leiurotoxin I is approximately 5 – 10-fold less potent in binding and biological activity than apamin (Chicchi et al. 1988).

Leiurotoxin I is a good example how an unknown toxin which constitutes only a very small fraction of the total venom protein, can be purified and pharmacologically characterized if the binding properties of a known toxin are used as a guideline. In this way, however, only other toxins of already known properties will be found.

4.3 Other Scorpion Toxins and Venoms

A scorpion toxin that affects both Na^+ and K^+ permeabilities in nerve axons was purified from the venom of the North African scorpion *Androctonus australis* Hector (Romey et al. 1975). The toxin, named neurotoxin I, blocked both Na^+ and K^+ conductances in *Sepia* giant axons. In contrast, in giant

Table 2. Scorpion toxins with specific effects on potassium currents and channels

Response which is reduced or abolished	Cell type	Toxin concentration	References
	Noxiustoxin		
Delayed outward K^+ current	Squid giant axon	$EC_{50} = 390$ nM	Carbone et al. 1982, 1987
Ca^{2+}-activated K^+ channels of high conductance	Skeletal muscle T-tubules of rat	$EC_{50} = 450$ nM	Valdivia et al. 1988
^{86}Rb efflux and GABA release	Synaptosomes from mouse brain	$EC_{50} \approx 3$ nM	Sitges et al. 1986
	Charybdotoxin		
Ca^{2+}-activated K^+ channels of high conductance	Skeletal muscle T-tubules of rat	$EC_{50} \approx 10$ nM	Miller et al. 1985; Smith et al. 1986; Anderson et al. 1988
Ca^{2+}-activated K^+ channels of high conductance	Apical cell membrane of nephron segments	≈ 5 nM	Guggino et al. 1987
Ca^{2+}-activated K^+ channels of intermediate conductance	Neurones of *Aplysia*	≈ 30 nM	Hermann and Erxleben 1987
Ca^{2+}-dependent K^+ current	Hippocampal CA1 neurones of rat	$\approx 20-30$ nM	Lancaster and Nicoll 1987; Storm 1987; Alger and Williamson 1988
Ca^{2+}-dependent K^+ current	Sympathetic ganglion cells of bullfrog	$\approx 4-20$ nM	Goh and Pennefather 1987
Ca^{2+}-activated K^+ channels	Rat pheochromocytoma (PC12) cells	≈ 10 µM	Hoshi and Aldrich 1988
Ca^{2+}-activated K^+ channels of high conductance	Primary bovine aortic smooth muscle	$EC_{50} = 2.1$ nM	Gimenez-Gallego et al. 1988
K^+ efflux induced by Ca-ionophore	GH$_3$ anterior pituitary cells	$EC_{50} = 2.1$ nM	Gimenez-Gallego et al. 1988
^{86}Rb efflux induced by Ca-ionophore	Human red cells	$IC_{50} = 0.9$ nM	Wolff et al. 1988
Ca^{2+}-dependent K^+ current	Cultured glioma C6 cells of rat	$IC_{50} = 1.6$ nM	Tas et al. 1988
Voltage-operated, Ca^{2+}-insensitive K^+ channels of low conductance	Motor nerve terminal of mouse	$3.5-30$ nM	Tabti et al. 1989
Transient K^+ current (A-current)	Murine T lymphocytes	5 nM	Lewis and Cahalan 1988
	Shaker K_A^+ channels from *Drosophila* expressed in *Xenopus* oocytes	$IC_{50} = 3.6$ nM	MacKinnon et al. 1988
	Leiurotoxin I		
Delayed rectifying K^+ current	Dorsal root ganglion cells of rat	$IC_{50} = 30$ nM	Schweitz et al. 1989
Delayed rectifying K^+ current	Frog node of Ranvier	$10-500$ nM	Bräu et al. 1990
K^+ efflux induced by angiotensin II	Hepatocytes of guinea pig	Partially purified toxin	Castle and Strong 1986
Adrenaline-induced relaxation	Taenia coli of guinea pig	$EC_{50} = 6.5$ nM	Chicchi et al. 1988

axons of crayfish and lobster the toxin inhibited the closing of open Na^+ channels, while it still blocked the K^+ currents. The effective concentration was fairly low with about 500 nM.

In frog myelinated nerve fibres (Pappone and Cahalan 1987) and in cultured rat anterior pituitary (GH_3) cells (Pappone and Lucero 1988) the crude venom of the scorpion *Pandinus imperator* blocked voltage-dependent K^+ currents in a voltage- and concentration-dependent manner. It has been claimed that in the nerve fibres *Pandinus* venom acts on the same component of the delayed outward current that is the target of dendrotoxin I (a component of a snake venom, see Sect. 5.1). The effect on the K^+ current is irreversible suggesting that the responsible toxin possesses a high affinity for K^+ channels which makes it an interesting candidate for purification.

5 Toxins from Snake Venoms

Snake venoms are sources of numerous substances including neurotoxins, cardiotoxins, myotoxins, coagulants and enzymes. This was found with snakes of the family Elapidae which includes the cobras, mambas and kraits, and of the family Crotalidae with the known rattle snakes. One class of proteins has attracted particular interest as they inhibit transmitter release from peripheral and central nerve terminals due to their intrinsic phospholipase A_2 (PLA_2) activity (for reviews see Chang 1979; Karlsson 1979; Howard and Gundersen 1980; Harris 1985). According to their different molecular structure they can

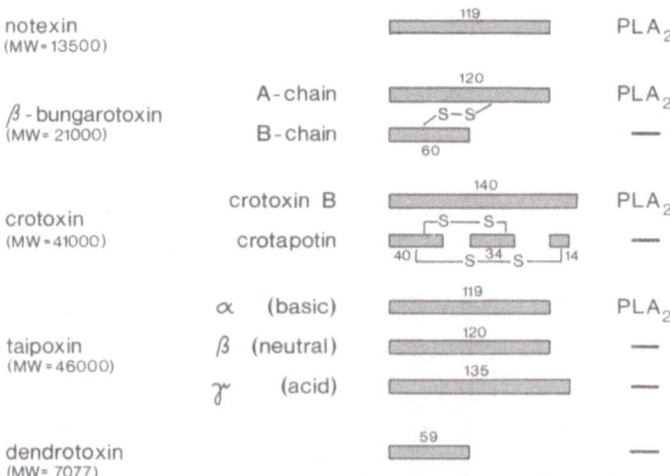

Fig. 3. Molecular structure of the most studied representatives of so called presynaptic-active neurotoxins. The schematic figure includes the molecular weight (MW), the number of amino acids of the peptide chains and indicates which peptide chains possess phospholipase A_2 activity (PLA_2)

be divided into four groups. The main and most studied representatives are shown in Fig. 3.

1. Notexin is a product of the Australian tiger snake *Notechis scutatus*. It is a basic, single chain protein with strong PLA_2 activity.
2. β-Bungarotoxin, a component of the Formosan banded krait *Bungarus multicinctus*, consists of two polypeptide chains covalently linked together by a disulfide bridge. The A-chain is a basic PLA_2, while the smaller B-chain has a high sequence homology to dendrotoxin and to certain protease inhibitors like the bovine pancreatic trypsin inhibitor.
3. Crotoxin from the South American rattle snake *Crotalus durissus terrificus* is a complex of two components which are not covalently linked. The larger chain, crotoxin B, is a basic, single-chain polypeptide with PLA_2 activity of only low lethal potency. The smaller acidic chain, called crotapotin or crotoxin A, consists of three much smaller, disulfide linked chains. Combination of both subunits increases tremendously the toxicity of phospholipase A_2 whereas crotapotin itself is non-toxic.
4. Taipoxin from the Australian taipan *Oxyuranus scutellatus* is a complex of three non-covalently linked subunits. Although all of them have a high homology to PLA_2, in fact only the α-subunit possesses PLA_2 activity. Together the three subunits yield the most potent snake toxin known.

These toxins described so far have in common that their blocking action on transmitter release is preceded by a facilitatory phase in which the amplitudes of nerve-evoked endplate potentials are increased. On the other hand there is another group of snake toxins that only facilitate evoked quantal release, but lack the final inhibition (Harvey and Karlsson 1980, 1982). The main representatives are dendrotoxin from the green mamba *Dendroaspis angusticeps* and toxin I (also often named dendrotoxin I) from the black mamba *Dendroaspis polylepis*. These toxins are basic, single-chain peptides, but most importantly, they are devoided of PLA_2 activity.

Generally, the PLA_2-toxins as well as the dendrotoxins inhibit K^+ currents that may be responsible for their facilitatory action at neuromuscular junctions. However, they can no longer be regarded as toxins acting exclusively on the transmitter release, because they also block K^+ currents in non-synaptic nerve membranes.

5.1 Dendrotoxins

First indications that the venom from the Eastern green mamba *Dendroaspis angusticeps* contains a peptide toxin which facilitates neuromuscular transmission came from studies of Barret and Harvey (1979). Soon after Harvey and Karlsson (1980) separated the venom into several components of which one polypeptide, named dendrotoxin, gained particular interest because it enhanced the quantal acetylcholine release at vertebrate neuromuscular junctions (Harvey and Karlsson 1980; Harvey and Gage 1981). Dendrotoxin constitutes about 2.5% of the total venom protein. It is composed of 59 amino acids with a molecular mass of 7077 (Harvey and Karlsson 1980; Joubert and Taljaard 1980). Two similar polypeptides, called toxin I and K, have been isolated from the venom of the black mamba *Dendroaspis polylepis* (Strydom 1973). Toxin I is the main protein of the venom with a yield of about 8%, while toxin K contributes about 3%. They consist of 60 and 57 amino acid residues with molecular weights of 7573 and 6542, respectively. Like dendrotoxin these compounds facilitate quantal acetylcholine release. In keeping with this they exhibit a high degree of sequence homology with dendrotoxin, differing only by 6 and 20 amino acids, respectively (Harvey and Karlsson 1982). Although all these dendrotoxins have a high degree of sequence homology with several protease inhibitors like the bovine pancreatic trypsin inhibitor (BPTI), the latter ones do not share their pharmacological properties (Harvey and Karlsson 1982; Anderson 1985; Dreyer and Penner 1987). A detailed comparison of members of the proteinase inhibitor family with four presynaptically active mamba snake toxins in respect to their amino acid sequences and structure-activity relation has been presented by Dufton (1985).

Reviews of Harvey et al. (1984b) and Harvey and Anderson (1985) deal extensively with the pharmacological effects and toxicities of mamba venoms, the purification and classification of the various peptide toxins, the chemical characterization of these toxins, structure-activity relationships and action on different nerve-muscle preparations.

From the venom of the green mamba Benishin et al. (1988) isolated four different polypeptides, one of them identical with dendrotoxin, the others of similar molecular mass (≈ 7000) and structure. These toxins are probably specific to different kinds of K^+ channels. However, they have not yet been studied in such detail as dendrotoxin and dendrotoxin I (toxin I) which deserve a more extended description.

5.1.1 General Effects

Dendrotoxins applied intravenously (Othman et al. 1982) or intraperitoneally (Velluti et al. 1987; Silveira et al. 1988) produce hyperactivity and induce typical convulsive symptoms in mice. In keeping with this an increased excitability of hippocampal neurones in vitro has been reported (Dolly et al. 1984). The toxins are much less toxic (about 10-fold) than the whole venom itself. Dendrotoxin and dendrotoxin I have LD_{50}-values of more than 20 mg/kg (mouse) following subcutaneous or intravenous administration (Strydom 1973; Joubert and Taljaard 1980). As with apamin (Sect. 3.1) their neurotoxicity is increased by more than thousandfold when injected intraventricularly into rat brain (Mehraban et al. 1985). Dendrotoxin I is always about a factor 5 more potent than dendrotoxin.

Dendrotoxins applied to chick biventer cervicis nerve-muscle preparation or mouse phrenic nerve-hemidiaphragm augment the nerve-induced twitch height if the bath temperature is above 30°C (Harvey and Karlsson 1980, 1982). Thus these preparations can be used as a convenient test of the toxin's biological activity. The minimal effective concentrations, however, are fairly high, i.e. 1 µg/ml.

The effects of dendrotoxin and dendrotoxin I have also been studied on peripheral autonomic neurotransmission (Harvey et al. 1984a; Anderson 1985) in mouse and rat vas deferens (predominantly sympathetic) and in chick oesophagus (parasympathetic). Both toxins augment the smooth muscle contractions in response to nerve stimulation, but once again the minimal effective toxin concentrations were found to be above 1 µM.

5.1.2 Binding of Dendrotoxins

Binding studies with dendrotoxins are often reported jointly with those performed with β-bungarotoxin and MCD peptide. Initial evidence that dendrotoxin and β-bungarotoxin share common binding sites in different preparations have been provided by Harvey and Karlsson (1982), demonstrating that dendrotoxin can antagonize the blocking effects of β-bungarotoxin on acetylcholine release in chick neuromuscular junctions, and by Othman et al. (1982), studying the binding of tritiated β-bungarotoxin to rat cortex synaptosomes and its inhibition by dendrotoxin I. Using ^{125}I-labelled dendrotoxin the presence of high-affinity ($K_D \approx 0.4$ nM) acceptor proteins for dendrotoxin on synaptosomal membranes of rat cortex (Mehraban et al. 1984; Black et al. 1986) and chick brain (Black and Dolly 1986) has been shown. The binding to these acceptors that are thought to be associated with K^+ channel proteins is inhibited by homologous polypeptides isolated from mamba snake venoms of *Dendroaspis polylepis* and *Dendroaspis viridis*. While dendrotoxin, dendrotoxin I or MCD peptide completely inhibit the binding of radiolabelled β-bungarotoxin to synaptic membrane fractions of rat and chick brain (Othman et al. 1982; Rehm and Betz 1984; Stansfeld et al. 1987; Schmidt et al. 1988), β-bungarotoxin is relatively inefficient in inhibiting ^{125}I-dendrotoxin binding to its high-affinity acceptors. The difference suggests the existence of at least two distinct acceptor proteins for dendrotoxin in synaptic membrane distinguishable by β-bungarotoxin (Black et al. 1986; Black and Dolly 1986; Stansfeld et al. 1987; Rehm and Lazdunski 1988b). In accordance with this the maximal binding capacity measured for ^{3}H-β-bungarotoxin is about sevenfold lower than that for ^{125}I-dendrotoxin.

As dendrotoxin has a potent and selective action on the delayed rectifying K^+ current in frog nerve fibres (see Sect. 5.1.3) Bräu et al. (1990) studied the binding of ^{125}I-dendrotoxin to the same preparation. The binding of dendrotoxin to the peripheral nerves ($K_D = 22$ pM) shows a 20-fold higher acceptor affinity than that to brain membranes, while the maximum binding capacity ($B_{max} \approx 23$ fmol/mg proteins) is about 30 to 50-fold lower than that determined for brain tissue. Binding of radiolabelled dendrotoxin could be displaced completely by dendrotoxin I ($K_I = 14$ pM) and MCD peptide ($K_I = 1.8$ nM). Remarkably, β-bungarotoxin ($K_I = 0.64$ nM) could inhibit only about 50% of the binding of labelled dendrotoxin. The binding of dendrotoxin was not affected by TEA, 3,4-DAP nor by capsaicin that blocks specifically the dendrotoxin-insensitive component of the delayed outward K^+ current in frog nerve fibres (Dubois 1982). Bräu et al. (1990) also applied immunocytochemical methods to visualize dendrotoxin-sensitive K^+ channels in the nodal and paranodal region of frog nerve fibres. Similar results, but with much higher binding capacity were obtained with rat nerve fibres (H. Repp and F. Dreyer, unpublished data). The advantage of peripheral nerve tissue in comparison to brain membranes which have so far been used in binding studies, is that the former are equipped with a well characterized population of K^+ channels.

Cross-linking of ^{125}I-labelled dendrotoxin (Mehraban et al. 1984; Black and Dolly 1986), toxin I and MCD peptide (Rehm et al. 1988) to their acceptors in rat and chick membranes and subsequent analysis by SDS-gel electrophoresis consistently reveals a polypeptide of the same apparent size of 76 kDa that has been solubilized and purified. This suggests that at least the snake toxins and the bee venom toxin bind to the same protein complex (Rehm et al. 1988; Rehm and Lazdunski 1988a), but they seem to bind to different sites (Bidard et al. 1987b; Schmidt et al. 1988). Further investigations of the binding protein have revealed that the dendrotoxin-sensitive K^+ channel is an oligomeric protein complex of 450000 Da (Black et al. 1988) consisting of polypeptide chains of 76000 Da and 38000 Da (Rehm and Lazdunski 1988a). Further purification analysis provided evidences for the existence of different populations of dendrotoxin I and MCD peptide binding proteins (Rehm and Lazdunski 1988b). The 76 kDa protein is of comparable mass to those that have been expressed in *Xenopus* oocytes from rat brain cDNA (Baumann et al. 1988) forming functional, dendrotoxin- and MCD peptide-sensitive, voltage-operated K^+ channels (Stühmer et al. 1988).

5.1.3 Dendrotoxin Blocks Selectively Delayed Rectifying Potassium Currents

The interest in dendrotoxin has been stimulated by the discovery that the toxin induces epileptiform activity and blocks a transient, voltage-operated K^+ current in hippocampal CA1 neurones while leaving other types of K^+ currents unaffected (Dolly et al. 1984; Halliwell et al. 1986).

Further studies on various types of nerve cell membranes revealed that dendrotoxin is more specific for the delayed rectifying K^+ current than for a transient outward K^+ current (Table 3). In myelinated frog nerve fibres the toxin prolongs remarkably the nerve action potential (Weller et al. 1985). Under voltage-clamp conditions it reduces the amplitude of the delayed K^+ current with nanomolar concentrations. It does not affect the leakage current nor the amplitude or the kinetic behaviour of the Na^+ current. About 30 to 80% of the delayed K^+ current is dendrotoxin-sensitive which supported the idea that different subtypes of delayed rectifier K^+ channels may exist. As already demonstrated by Dubois (1981) the delayed K^+ current in frog nerve fibres is composed of at least three kinetically distinct currents, one (I_{Kf2}) of them selectively inhibited by capsaicin (Dubois 1982). Using dendrotoxin I from the black mamba Benoit and Dubois (1986) showed that another component, called I_{Kf1}, is selectively and potently blocked by this toxin ($IC_{50} = 0.4$ nM). The same specific action is obtained with dendrotoxin from the green mamba, yet this toxin is about 20 times less potent (Bräu et al. 1990). The dendrotoxin-sensitive K^+ current has the typical feature of becoming activated at membrane potentials around -60 mV.

Dorsal root ganglion cells of guinea pig exhibit three different types of K^+ currents, a transient outward current, a Ca^{2+}-dependent current and a delayed non-inactivating outward current. At concentrations of 0.14 to 1.4 nM dendrotoxin, if applied from the outside, selectively blocks a portion (maximal 50%) of the non-inactivating K^+ current, leaving the transient outward current unaffected (Penner et al. 1986). Once the action of dendrotoxin is complete the inhibition of the K^+ current cannot be reversed within 20–30 min of continuous washing. Higher concentrations, up to 1.4 µM, did not further reduce the delayed outward K^+ current. Thus there are probably two distinct sets of K^+ channels. In fact the current-voltage relations are quite different for the dendrotoxin-sensitive and -insensitive component. The dendrotoxin-sensitive current has a threshold of about -50 mV. Its amplitude increases almost linearly with voltage (ohmic behaviours). The dendrotoxin-resistant current increases non-linearly with membrane depolarization up to $+60$ mV. Experiments performed with K^+ channel blockers revealed that dendrotoxin affects a type of K^+ channels which is more specific for 3,4-diaminopyridine than for TEA.

Stansfeld et al. (1986, 1987) studied the action of dendrotoxins on isolated visceral sensory neurones of rat nodose ganglia. In a minor subpopulation (A-cells) the toxins induce repetitive firing during prolonged depolarization on current injection. The toxins block about 50% of an outward rectifying K^+ current that shows slow, but incomplete inactivation within several seconds. This reminds us to the findings obtained with dendrotoxin in frog nodes of Ranvier and dorsal root ganglion cells. This current is further characterized by its inhibition by 4-aminopyridine at concentrations as low as 10 to 30 µM, and its resistance to agents such as TEA (10 mM), barium (3 mM) or cesium (4 mM).

In giant axons of the marine annelid *Myxicola* dendrotoxin inhibits the delayed rectifier K^+ current completely with an IC_{50}-value of 150 nM (Schauf 1987). Interestingly, the toxin was found equally effective when applied externally or internally in contrast to dorsal root ganglion cells where it acts only from the outside (Penner et al. 1986). In the *Myxicola* preparation dendrotoxin also slows the inactivation of the Na^+ current without altering other Na^+ channel properties (Schauf 1987).

Benishin et al. (1988) studied the ^{86}Rb efflux from rat forebrain synaptosomes which obviously possess four physiologically and pharmacologically distinct types of K^+ channels. From the crude venom of the green mamba (*Dendroaspis angusticeps*) they isolated four different polypeptides, all of similar molecular weight and structure. Two of them, one identical to dendrotoxin, block preferentially 4-aminopyridine-sensitive, inactivating outward rectifier K^+ channels ($IC_{50} \approx 15$ nM). The other ones, whose amino acid sequences are so far only partially known, are selective for the non-inactivating rectifier K^+ channels. These observations, however, have to be confirmed in other tissues under voltage-clamp conditions.

As the increase of nerve-induced twitch height and of quantal acetylcholine release at neuromuscular junctions were the initial effects described for dendrotoxin (Harvey and Karlsson 1980, 1982), it was of interest to study the mechanism by which the alterations were brought about at motor endplates. A likely possibility is an inhibition of K^+ permeability which prolongs presynaptic depolarization and thereby increases the Ca^{2+} influx.

Some of the currents flowing through the membrane of motor nerve terminals upon nerve stimulation can be investigated by the subendothelial recording technique (for details see Mallart 1985; Penner and Dreyer 1986). Apart from the activation of the Na^+ current in the nodes of Ranvier, two different types of Ca^+ currents and three different types of K^+ currents can be recorded. Dendrotoxin up to 1.5 μM affects neither the Na^+ current nor the fast K^+ current (Dreyer and Penner 1987; Anderson and Harvey 1988) nor the Ca^{2+}-activated K^+ current (Dreyer and Penner 1987) that is unmasked after blockade of the fast K^+ current by 3,4-diaminopyridine. Only a third slowly activating, 3,4-diaminopyridine-sensitive K^+ current that is revealed by inhibition of the other K^+ currents by TEA (30 mM), is blocked by dendrotoxin in a concentration-dependent manner (1.5–150 nM). Identical results were obtained with the PLA_2 snake toxins such as β-bungarotoxin (Sect. 5.2), taipoxin and crotoxin (Sect. 5.3), but the effective concentrations were three to four times higher (Dreyer and Penner 1987). In contrast, bee venom phospholipase A_2 (1 μg/ml) and the protease inhibitor aprotinin (1.5 μM) were found devoid of noticeable effects on the described presynaptic currents.

Dendrotoxin (3.9 μM) consistently causes repetitive firing of action potentials following single nerve stimulation that correlates with the occurrence of multiple endplate potentials (Anderson and Harvey 1988). The mechanism is most probably due to the blockade of the slowly activating K^+ current that has been characterized as dendrotoxin- and aminopyridine-sensitive (Dreyer and Penner 1987) which thus may have the function to prevent repetitive firing of nerve action potentials. The increase in quantal content of nerve-evoked endplate potentials (by 20−50%) with dendrotoxin is not maintained. It can not always be recorded neither in frog nor in mouse nerve-muscle preparations (Anderson and Harvey 1988). Therefore, the repetitive activity, but not the more or less pronounced increase in nerve induced quantal transmitter release, seems to be the main cause for the augmentation of twitch height in dendrotoxin-treated nerve-muscle preparations. This explains also why the maximal augmentation is always two or three times above control.

Dendrotoxin-sensitive K^+ channels obviously also exist in the axons and terminals of rat neurointermediate lobes. Here dendrotoxin ($EC_{50} \approx 2$ nM) like 4-aminopyridine ($EC_{50} \approx 2$ μM) augments the electrically evoked secretion of vasopressin and oxytocin (Bondy and Russel 1988).

5.1.4 Dendrotoxin-Sensitive Potassium Channels

From the initial description of an action of dendrotoxin on K^+ currents it took four years to identify the underlying K^+ channels on the single channel level. Stansfeld and Feltz (1988) described single channel properties of dendrotoxin-sensitive channels in dorsal root ganglion cells. In physiological K^+ gradient single channel records gave a slope conductance of 5−10 pS which increased to 18 pS (outward current) in symmetrical high-K^+ solution. These dendrotoxin-sensitive, voltage-operated K^+ channels are active at resting potential, show little inactivation within 500 ms, and are inhibited by 2 nM toxin. The action of the toxin is marked by disappearance of single channel currents rather than by flickering or reduction of the current amplitude. Stühmer et al. (1988) expressed in *Xenopus* oocytes an ion channel protein encoded by cDNA of rat cerebral cortex. It exhibits properties similar to those of the "classical" very slowly or non-inactivating delayed rectifier K^+ channel. In physiological solutions the single channel conductance is about 9 pS and it is increased to 22 pS in symmetrical 100 mM K^+ medium. Pharmacologically this K^+ channel is characterized by (1) its sensitivity to dendrotoxin ($IC_{50} \approx 12$ nM), MCD peptide ($IC_{50} \approx 45$ nM), TEA ($IC_{50} \approx 0.6$ mM) and 4-aminopyridine ($IC_{50} \approx 1.0$ mM), (2) its insensitivity to apamin (1 μM)or β-bungarotoxin (200 nM) (Stühmer et al. 1988). The pharmacological as well as the physiological properties are in best agreement with those of the I_{Kf1} current, a subtype of the delayed outward rectifying K^+ current described in frog node of Ranvier (Bräu et al. 1990). The underlying K^+ channels have recently been investigated by patch-clamp experiments on axonal membranes from the nodal and internodal region of frog myelinated nerve fibres (Jonas et al. 1989). In symmetrical high-K^+ solution the unitary conductance of the dendrotoxin-sensitive K^+ channel was 23 pS.

5.1.5 Action of Dendrotoxin on Cells of the Central Nervous System

Apart from the action of dendrotoxin on delayed outward K^+ currents in peripheral nerve tissue two reports deal with the effect of the toxin on K^+ currents in central neurones. In fact the transient, voltage-activated K^+ current (A-current) in hippocampal CA1 neurones was the first K^+ current that has been found to be blocked by dendrotoxin while other types of K^+ currents in these neurones remained unaffected (Dolly et al. 1984; Halliwell et al. 1986). In hippocampal slices, however, it takes 30 min treatment with 300 nM dendrotoxin to reduce the A-current by more than 70%. Toxin I was found about 5 times more potent.

Magnocellular neurosecretory neurones, located in the supraoptic nucleus of rat hypothalamus, display a fast transient, Ca^{2+}-dependent K^+ current which is distinct from the I_A current in vertebrate neurones. It is activated from a threshold of -75 mV and completely inactivated at potentials positive to -55 mV. This current is completely unaffected by TEA at concentrations up to 12 mM, but markedly sensitive to 4-aminopyridine ($IC_{50} \approx 1$ mM) and to dendrotoxin ($IC_{50} \approx 4$ nM) (Bourque 1988).

Apart from its central action, dendrotoxin did not suppress a transient, fast inactivating A-current in C cells of rat nodose ganglia and in neurones of rat superior cervical ganglia (Stansfeld et al. 1987), nor the transient K^+ current in guinea pig dorsal root ganglion cells (Penner et al. 1986; Stansfeld and Feltz 1988) and in mammalian motor nerve terminals (Dreyer and Penner 1987).

5.1.6 Properties of the Dendrotoxin-Sensitive Potassium Channel: a Summary

From available electrophysiological data the properties of the dendrotoxin-sensitive K^+ current and its underlying K^+ channels may at present be summarized as follows:

1. The dendrotoxin-sensitive K^+ channel most probably exists only in neuronal membranes.
2. Although in central neurones effects on a transient, Ca^{2+}-activated K^+ current and on A-current have been observed, the main target of dendrotoxin action is a subtype of the slowly or non-inactivating outward rectifier K^+ channel.

3. The activation of the K^+ channel is steeply voltage-dependent starting from a threshold varying between -70 mV and -30 mV for different types of cells. The current-voltage relationship is almost linear for more positive membrane potentials.

4. The unitary conductance of the dendrotoxin-sensitive K^+ channel is about 10 pS and 20 pS in physiological K^+ gradient and in symmetrical high-K^+ solutions, respectively.

5. The primary structure of this K^+ channel has been reported. The channel molecule seems to be formed by four subunits consisting of 495 amino acid residues each.

6. *All* dendrotoxin-sensitive K^+ channels are inhibited by aminopyridines, often with higher potency and, noteworthy, higher efficacy compared to TEA.

7. While apamin is without effect, MCD peptide blocks the dendrotoxin-sensitive K^+ channel with the same efficacy, but with a 5 to 10 times lower potency. It is more or less affected by snake toxins with phospholipase A_2 activity depending strongly on the cell type.

5.1.7 Cell Preparations Where Dendrotoxin is Ineffective

To further define the specificity of dendrotoxin the toxin was applied to a variety of excitable and non-excitable cell preparations where different types of K^+ currents and channels can be studied (unpublished data). No toxin effect was observed in (1) smooth muscles of guinea pig bladder (G. Isenberg, U. Klöckner, F. Dreyer), (2) atrial and ventricular cells of guinea pig heart (G. Isenberg, U. Klöckner, F. Dreyer), (3) bovine adrenal chromaffin cells (F. Dreyer), (4) chick embryonic fibroblasts (F. Dreyer and G. Seidel), (5) rat anterior pituitary GH_3 tumor cells (G. Oxford and W. Vogel) and (6) neuroblastoma cells of clone Neuro2-A and N1E 115 (F. Dreyer). It should be mentioned that these neuroblastoma cells do not bind tetanus toxin, which is assumed to be a specific marker of neuronal membranes.

5.2 Beta-Bungarotoxin

In the past β-bungarotoxin has aroused some interest because it inhibits transmitter release from peripheral and central nerve terminals. Using radiolabelled toxin, high-affinity binding to protein acceptors ($K_D \approx 0.5$ nM) on rat and chick brain membranes has been shown (Oberg and Kelly 1976; Othman et al. 1982; Rehm and Betz 1982), and the purified acceptors were solubilized and characterized (Rehm and Betz 1984; Black and Dolly 1986; Black et al. 1988; Rehm and Lazdunski 1988b). The results have already been mentioned in Sects. 3.2.1 and 5.1.2 in connection with the binding of MCD peptide and dendrotoxin, respectively. The binding experiments demonstrated that the binding of β-bungarotoxin on synaptic membrane fractions is completely inhibited by dendrotoxin and MCD peptide (Othman et al. 1982; Rehm and Betz 1984; Stansfeld et al. 1987; Schmidt et al. 1988; Schmidt and Betz 1988), but in reverse β-bungarotoxin can only partially prevent the binding of dendrotoxin and MCD peptide suggesting the existence of two subpopulations of dendrotoxin acceptor sites (Black et al. 1986; Black and Dolly 1986; Rehm and Lazdunski 1988b; Schmidt et al. 1988).

The β-bungarotoxin binding site is part of a large membrane protein complex (molecular mass $\approx 430\,000$) (Rehm and Betz 1984) that presumably consists of several polypeptide subunits. One of them, with a molecular mass of $95\,000$, seems to be the toxin binding site (Rehm and Betz 1983).

β-Bungarotoxin binding sites, according to autoradiographic studies in rat brain are prominent in cerebral cortex, parts of the hippocampus and in the molecular layer of cerebellum. Much less binding is found in various other regions (Othman et al. 1983; Pelchen-Matthews and Dolly 1988). Very little binding is associated with myelinated tracts. β-Bungarotoxin binding sites are not exclusively localized on nerve terminals because toxin acceptors have also been identified on ganglion cells which had lost their presynaptic endings (Rehm and Betz 1982) and on peripheral myelinated frog and mammalian nerve fibres (Bräu et al. 1990; H. Repp and F. Dreyer, unpublished data).

5.2.1 Beta-Bungarotoxin Blocks Voltage-Operated Potassium Currents

Direct evidence that β-bungarotoxin blocks, like dendrotoxin, K^+ channels has first been provided by Petersen et al. (1986) in guinea pig dorsal root ganglion cells and by Dreyer and Penner (1987) in mouse motor nerve terminals. In dorsal root ganglion cells β-bungarotoxin inhibits a portion of the non-inactivating outward K^+ current. However, additional application of dendrotoxin further reduced this outward rectifier (Petersen et al. 1986). Thus in this preparation β-bungarotoxin is much less potent than dendrotoxin and also less efficient (Table 3). In myelinated frog nerve fibres the dendrotoxin-sensitive delayed outward K^+ current is also blocked by β-bungarotoxin ($IC_{50} = 818$ nM) (Bräu et al. 1990).

A potent, concentration-dependent inhibitory effect of β-bungarotoxin ($5-500$ nM) on a slowly activating, aminopyridine-sensitive, but less TEA-sensitive K^+ current has been demonstrated in mouse motor nerve terminals with the subendothelial recording technique (Dreyer and Penner 1987). The toxin (500 nM) suppressed neither the Na^+ current nor the fast voltage-activated nor the Ca^{2+}-activated K^+ current components. Rowan and Harvey (1988), who studied the fast K^+ component of the nerve terminal membrane current, did observe an approximate 25% reduction in mammalian nerve terminals, but no effect in frog nerve endings.

The action of β-bungarotoxin and related PLA$_2$ toxins (see Sect. 5.3) is unrelated to the enzymatic activity as (1) the bee venom phospholipase A$_2$, despite of its higher enzymatic activity compared to the toxins, did not affect K$^+$ currents and (2) the inhibition of K$^+$ currents is still observed at 20°C at which temperature the PLA$_2$ activity is less than 5% compared to that at 37°C.

5.3 Crotoxin, Taipoxin, Notexin

The presynaptically active neurotoxins with PLA$_2$ activity differ considerably in their molecular structure (Fig. 3). This suggests that they may act through different receptors. Indeed, some indirect evidence for this came from double-poisoning experiments with dendrotoxin, β-bungarotoxin, crotoxin, taipoxin and notexin (Chang and Su 1980; Harvey and Karlsson 1982). The results were confirmed by binding experiments with ^{125}I-labelled neurotoxins to synaptosomal membrane preparations from guinea pig brain (Tzeng et al. 1986). While the binding of radiolabelled β-bungarotoxin, crotoxin and taipoxin is completely inhibited by their unlabelled molecules with IC$_{50}$-values of 11, 0.51 and 45 nM, respectively, these neurotoxins do not significantly compete with each other for binding to their acceptors. This suggests that in synaptosomal membranes separate binding sites of high-affinity exist for the three toxins. Crotoxin is also unable to inhibit the binding of dendrotoxin to chick synaptosomes (Black and Dolly 1986). It should be noted that in contrast to β-bungarotoxin that obviously acts selectively on neuronal cells, both crotoxin and taipoxin display additional specific binding to liver, kidney, lung and erythrocyte ghosts. Using the photoaffinity labelling technique Tzeng et al. (1986) identified for crotoxin an acceptor protein of about 85 kDa in brain membranes.

Like dendrotoxin and β-bungarotoxin, crotoxin and taipoxin affect an aminopyridine-sensitive, slowly activating K$^+$ current in mouse motor nerve terminals (Dreyer and Penner 1987). Rowan and Harvey (1988) tested all four phospholipase toxins shown in Fig. 3 on mouse motor nerve terminals and found a reduction of a K$^+$ current by about 50% (Table 3). The binding experiments of Tzeng et al. (1986) suggest that the toxins may affect the K$^+$ channels by binding to different sites of the same channel protein complex. Alternatively the toxins might share the same binding site concerning the block of K$^+$ channel, but might have quite different acceptor sites responsible for their enzymatic activity on cell membranes.

Table 3. Blockade of potassium currents and channels by dendrotoxin (DTX), Dendrotoxin I (DTX I), mast cell degranulating peptide (MCDP), β-bungarotoxin (β-BuTX), Crotoxin (CroTX), taipoxin (TPX) and notexin (NoTX)

Response which is reduced or abolished	Cell preparation	Toxin	Concentration	References
Transient outward K^+ current (I_A)	CA1 neurones in hippocampal slices from rat and guinea pig	DTX	50–300 nM	Dolly et al. 1984;
		DTX I	10–150 nM	Halliwell et al. 1986
Delayed outward rectifier I_K	Frog node of Ranvier	DTX	0.1–80 nM	Weller et al. 1985
Delayed outward rectifier I_K (component I_{Kf1})	Frog node of Ranvier	DTX I	$IC_{50} = 0.4$ nM	Benoit and Dubois 1986
		DTX	$IC_{50} = 11$ nM	Bräu et al. 1990
		MCDP	$IC_{50} = 33$ nM	
		β-BuTX	$IC_{50} = 940$ nM	
Delayed rectifying K^+ current	Giant axon of marine annelid Myxicola	DTX	$IC_{50} = 150$ nM	Schauf 1987
Delayed outward rectifier I_K (one component)	Dorsal root ganglion cells of guinea pig	DTX	0.14–1.4 nM	Penner et al. 1986
Outward rectifying K^+ current	A-cells of rat nodose ganglia	β-BuTX	0.45–45 nM	Petersen et al. 1986
		DTX	$IC_{50} = 2.1$ nM	Stansfeld et al. 1986, 1987
		MCDP	$IC_{50} = 37$ nM	
		DTX I	max. 10 nM	
Slow activating K^+ current	Motor nerve terminals of mouse	DTX	1.5–150 nM	Dreyer and Penner 1987
		β-BuTX	5–500 nM	
		CroTX	5–500 nM	
		TPX	5–500 nM	
		MCDP	3–300 nM	Beise and Dreyer (unpublished data)
K^+ current	Motor nerve terminals of mouse	β-BuTX	≈ 150 nM	Rowan and Harvey 1988
		CroTX	≈ 130 nM	
		TPX	≈ 6.5 nM	
		NoTX	≈ 250 nM	
^{86}Rb efflux through inactivating K^+ channels	Forebrain synaptosomes of rat	DTX	$IC_{50} \approx 15$ nM	Benishin et al. 1988
Electrically evoked secretion of vasopressin and oxytocin	Neurohypophysial neurones of rat	DTX	$EC_{50} \approx 2$ nM	Bondy and Russel 1988
Transient Ca^{2+}-dependent K^+ current	Magnocellular neurosecretory cells of rat hypothalamus	DTX	$IC_{50} \approx 4$ nM	Bourque 1988
Delayed rectifier K^+ channel (5–10 pS)	Dorsal root ganglion cells of rat	DTX	≈ 2 nM	Stansfeld and Feltz 1988
Delayed rectifier K^+ channel (9 pS)	Rat brain cDNA expressed in Xenopus oocytes	DTX	$IC_{50} = 12$ nM	Stühmer et al. 1988
		MCDP	$IC_{50} = 45$ nM	
Delayed rectifier K^+ channel (23 pS in high K^+ solution)	Nodal and internodal membrane of frog myelinated nerve fibres	DTX	$EC_{50} \approx 50$ nM	Jonas et al. 1989

6 Toxins from Cone Snails

The venoms of the fish-eating gastropod cone snails which are spread over the Australian and Pacific Area, are rich sources of different biologically active peptides. These colourful, beautifully patterned animals which might appear perfectly harmless can be lethal for man. There are numerous species and so far from only a few of them peptide neurotoxins have been isolated, sequenced and biologically characterized. Among these conotoxins, specific blockers of nicotinic acetylcholine receptors (α-conotoxin), of voltage-activated Na^+ channels in muscle membrane (μ-conotoxin) and of a subtype of vertebrate neuronal voltage-sensitive Ca^{2+} channels (ω-conotoxin) were found (for reviews see Cruz et al. 1985; Olivera et al. 1985; Gray et al. 1988).

First hints that the cone snail venoms may also contain components specific for K^+ channels, came from the work of Chesnut et al. (1987). They studied the effect of crude venom, extracted from the venom ducts of the species *Conus striatus*, on the delayed rectifier current in neurones of *Aplysia californica*. The venom reduced, or increased in a concentration dependent manner the K^+ current depending on the type of neuron. Regardless in all cells it slowed both the activation and inactivation of the current. As some differentiation in the venom effect could be observed after heat treatment and a filtration procedure (using a cutoff of 50 000 molecular mass) it was suggested that the various effects on the K^+ current are caused by different components of the venom. This conclusion should be regarded as preliminary and ought to be confirmed with purified peptide components. In view of the variety of species still to be explored we can hope that the horn of plenty contains at least some toxins highly specific for K^+ channels, thus making systematic studies for such conotoxins worthwhile.

7 Concluding Remarks

In the past potent biological toxins such as tetrodotoxin and α-bungarotoxin have been important in the isolation and characterization of ion channel protein complexes such as the sodium channel and the nicotinic acetylcholine receptor. The K^+ channel peptide toxins have never played a comparable role due to the fast development of gentechnology and molecular biology which now allows the biochemical characterization of ion channels without any specific ligand. The toxins may allow a simple classification of the K^+ channels. This is highly desirable particularly since the patch clamp technique has revealed already so many kinds of K^+ channels. Classification based only on single channel conductance and kinetic behaviour becomes more and more difficult.

Furthermore, excitable as well as non-excitable cells possess generally several types of K^+ channels causing complex membrane currents that are often difficult to separate. K^+ channel toxins as highly specific probes can help to dissect complex K^+ conductance changes and to evaluate the physiological role of the different K^+ channels and their currents in the function and activity of developing and mature cells.

Many papers dealing with the action of toxins start or end with remarks like "toxins are important tools for biochemists and electrophysiologists to study structure and function of drug and voltage-operated ion channels". So far only few toxins got this approval. To reach this level the specificity of the toxin's action must be known in detail. Only broad application in different species and cell types will finally decide whether the toxin gains permanent value as a tool or not.

The past 10 years have provided us with some peptide toxins that affect some types of the large family of K^+ channels with high specificity and affinity, but there are still many other family members still waiting for their lids. In this respect it should be kept in mind that putative toxins are mostly found by luck rather than by systematic research. This means that at present we are still in the age of hunters and gatherers. However, I am sure that the many venomous animals — which are less dangerous than mankind itself — can offer us further peptide toxins to close the gap which still exists for the K^+ channel family.

Acknowledgements. I am very grateful to Drs. E. Habermann and Chr. Walther for much helpful comments on the manuscript.

References

Abia A, Lobaton CD, Moreno A, Garcia-Sancho J (1986) *Leiurus quinquestriatus* venom inhibits different kinds of Ca^{2+}-dependent K^+ channels. Biochim Biophys Acta 856:403–407

Alger BE, Williamson A (1988) A transient calcium-dependent potassium component of the epileptiform burst after-hyperpolarization in rat hippocampus. J Physiol (Lond) 399:191–205

Anderson AJ (1985) The effects of protease inhibitor homologues from mamba snake venoms on autonomic neurotransmission. Toxicon 23:947–954

Anderson AJ, Harvey AL (1988) Effects of the potassium channel blocking dendrotoxins on acetylcholine release and motor nerve terminal activity. Br J Pharmacol 93:215–221

Anderson CS, MacKinnon R, Smith C, Miller C (1988) Charybdotoxin block of single Ca^{2+}-activated K^+ channels. Effects of channel gating, voltage, and ionic strength. J Gen Physiol 91:317–333

Banks BEC, Brown C, Burgess GM, Burnstock G, Claret M, Cocks TM, Jenkinson DH (1979) Apamin blocks certain neurotransmitter-induced increases in potassium permeability. Nature 282:415–417

Barrett JC, Harvey AL (1979) Effects of the venom of the green mamba, *Dendroaspis angusticeps* on skeletal muscle and neuromuscular transmission. Br J Pharmacol 67: 199–205

Baumann A, Grupe A, Ackermann A, Pongs O (1988) Structure of the voltage-dependent potassium channel is highly conserved from *Drosophila* to vertebrate central nervous system. EMBO J 7:2457–2463

Benishin CG, Sorensen RG, Brown WE, Krueger BK, Blaustein MP (1988) Four polypeptide components of green mamba venom selectively block certain potassium channels in rat brain synaptosomes. Mol Pharmacol 34:152–159

Benoit E, Dubois J-M (1986) Toxin I from the snake *Dendroaspis polylepis polylepis*: a highly specific blocker of one type of potassium channel in myelinated nerve fiber. Brain Res 377:374–377

Bidard J-N, Gandolfo G, Mourre C, Gottesmann C, Lazdunski M (1987a) The brain response to the bee venom peptide MCD. Activation and desensitization of a hippocampal target. Brain Res 418:235–244

Bidard J-N, Mourre C, Lazdunski M (1987b) Two potent central convulsant peptides, a bee venom toxin, the MCD peptide, and a snake venom toxin, dendrotoxin I, known to block K⁺ channels, have interacting receptor sites. Biochem Biophys Res Commun 143:383–389

Black AR, Dolly JO (1986) Two acceptor sub-types for dendrotoxin in chick synaptic membranes distinguishable by β-bungarotoxin. Eur J Biochem 156:609–617

Black AR, Breeze AL, Othman IB, Dolly JO (1986) Involvement of neuronal acceptors for dendrotoxin in its convulsive action in rat brain. Biochem J 237:397–404

Black AR, Donegan CM, Denny BJ, Dolly JO (1988) Solubilization and physical characterization of acceptors for dendrotoxin and β-bungarotoxin from synaptic membranes of rat brain. Biochemistry 27:6814–6820

Blatz AL, Magleby KL (1986) Single apamin-blocked Ca-activated K⁺ channels of small conductance in cultured rat skeletal muscle. Nature 323:718–720

Bondy CA, Russel JT (1988) Dendrotoxin and 4-aminopyridine potentiate neurohypophysial hormone secretion during low frequency electrical stimulation. Brain Res 453:397–400

Bourque CW (1988) Transient calcium-dependent potassium current in magnocellular neurosecretory cells of the rat supraoptic nucleus. J Physiol (Lond) 397:331–347

Bourque CW, Brown DA (1987) Apamin and d-tubocurarine block the afterhyperpolarization of rat supraoptic neurosecretory neurons. Neurosci Lett 82:185–190

Bräu ME, Dreyer F, Jonas P, Repp H, Vogel W (1990) A K⁺ channel in *Xenopus* nerve fibres selectively blocked by bee and snake toxins: binding and voltage-clamp experiments. J Physiol (Lond) 420:365–385

Breithaupt H, Habermann E (1968) Mastzelldegranulierendes Peptid (MCD-Peptid) aus Bienengift: Isolierung, biochemische und pharmakologische Eigenschaften. Naunyn-Schmiedeberg's Arch Pharmacol 261:252–270

Brown PD, Sepúlveda FV (1985) Potassium movements associated with amino acid and sugar transport in enterocytes from rabbit jejunum. J Physiol (Lond) 363:271–285

Burgess GM, Claret M, Jenkinson DH (1981) Effects of quinine and apamin on the calcium-dependent potassium permeability of mammalian hepatocytes and red cells. J Physiol (Lond) 317:67–90

Carbone E, Wanke E, Prestipino G, Possani LD, Maelicke A (1982) Selective blockage of voltage-dependent K⁺ channels by a novel scorpion toxin. Nature 296:90–91

Carbone E, Prestipino G, Spadavecchia L, Franciolini F, Possani LD (1987) Blocking of the squid axon K⁺ channel by noxiustoxin: a toxin from the venom of the scorpion *Centruroides noxius*. Pflügers Arch 408:423–431

Castle NA, Strong PN (1986) Identification of two toxins from scorpion (*Leiurus quinquestriatus*) venom which block distinct classes of calcium-activated potassium channel. FEBS Lett 209:117–121

Castle NA, Haylett DG, Jenkinson DH (1989) Toxins in the characterization of potassium channels. Trends Neurosci 12:59–65

Chang CC (1979) The action of snake venoms on nerve and muscle. In: Lee CY (ed) Snake venoms, chap 10. Springer, Berlin Heidelberg New York, pp 309–376 (Handbook of experimental pharmacology, vol 52)

Chang CC, Su MJ (1980) Mutual potentiation, at nerve terminals, between toxins from snake venoms which contain phospholipase A activity: β-bungarotoxin, crotoxin, taipoxin. Toxicon 18:641–648

Cherubini E, Ben Ari Y, Gho M, Bidard JN, Lazdunski M (1987) Long-term potentiation of synaptic transmission in the hippocampus induced by a bee venom peptide. Nature 328:70–73

Cherubini E, Neuman R, Rovira C, Ben Ari Y (1988) Epileptogenic properties of the mast cell degranulating peptide in CA3 hippocampal neurones. Brain Res 445:91–100

Chesnut TJ, Carpenter DO, Strichartz GR (1987) Effects of venom from *Conus striatus* on the delayed rectifier potassium current of molluscan neurons. Toxicon 25:267–278

Chicchi GG, Gimenez-Gallego G, Ber E, Garcia ML, Winquist R, Cascieri MA (1988) Purification and characterization of a unique, potent inhibitor of apamin binding from *Leiurus quinquestriatus hebraeus* venom. J Biol Chem 263:10192–10197

Cognard C, Traoré F, Potreau D, Raymond G (1984) Effects of apamin on the outward potassium current of isolated frog skeletal muscle fibres. Pflügers Arch 402:222–224

Cook NS (1988) The pharmacology of potassium channels and their therapeutic potential. Trends Pharmacol Sci 9:21–28

Cook NS, Haylett DG (1985) Effects of apamin, quinine and neuromuscular blockers on calcium-activated potassium channels in guinea-pig hepatocytes. J Physiol (Lond) 358:373–394

Cook NS, Haylett DG, Strong PN (1983) High affinity binding of [^{125}I]monoiodoapamin to isolated guinea-pig hepatocytes. FEBS Lett 152:265–269

Costa M, Furness JB, Humphreys CMS (1986) Apamin distinguishes two types of relaxation mediated by enteric nerves in the guinea-pig gastrointestinal tract. Naunyn-Schmiedeberg's Arch Pharmacol 332:79–88

Cruz LJ, Gray WR, Yoshikami D, Olivera BM (1985) Conus venoms: a rich source of neuroactive peptides. J Toxicol-Toxin Rev 4:107–132

Den Hertog A (1981) Calcium and the alpha-action of catecholamines on guinea-pig taenia caeci. J Physiol (Lond) 316:109–125

Dolly JO, Halliwell JV, Black JD, Williams RS, Pelchen-Matthews A, Breeze AL, Mahraban F, Othman IB, Black AR (1984) Botulinum neurotoxin and dendrotoxin as probes for studies on transmitter release. J Physiol (Paris) 79:280–303

Dreyer F, Penner R (1987) The actions of presynaptic snake toxins on membrane currents of mouse motor nerve terminals. J Physiol (Lond) 386:455–463

Dubois J-M (1981) Evidence for the existence of three types of potassium channels in the frog Ranvier node membrane. J Physiol (Lond) 318:297–316

Dubois JM (1982) Capsaicin blocks one class of K$^+$ channels in the frog node of Ranvier. Brain Res. 245:372–375

Dufton MJ (1985) Proteinase inhibitors and dendrotoxins. Sequence classification, structural prediction and structure/activity. Eur J Biochem 153:647–654

Field A, Jenkinson DH (1987) The effect of noradrenaline on the ion permeability of isolated mammalian hepatocytes, studied by intracellular recording. J Physiol (Lond) 392:493–512

Fosset M, Schmid-Antomarchi H, Hugues M, Romey G, Lazdunski M (1984) The presence in pig brain of an endogenous equivalent of apamin, the bee venom peptide that specifically blocks Ca^{2+}-dependent K$^+$ channels. Proc Natl Acad Sci USA 81:7228–7232

Galvan M, Behrends J (1985) Apamin blocks calcium-dependent spike after-hyperpolarization in rat sympathetic neurones. Pflügers Arch 403:R50

Gater PR, Haylett DG, Jenkinson DH (1985) Neuromuscular blocking agents inhibit receptor-mediated increases in the potassium permeability of intestinal smooth muscle. Br J Pharmacol 86:861–868

Gimenez-Gallego G, Navia MA, Reuben JP, Katz GM, Kaczorowski GJ, Garcia ML (1988) Purification, sequence, and model structure of charybdotoxin, a potent selective inhibitor of calcium-activated potassium channels. Proc Natl Acad Sci USA 85:3329–3333

Goedert M, Hunter JC, Ninkovic M (1984) Evidence for neurotensin as a non-adrenergic, non-cholinergic neurotransmitter in guinea pig ileum. Nature 311:59–62

Goh JW, Pennefather PS (1987) Pharmacological and physiological properties of the after-hyperpolarization current of bullfrog ganglion neurones. J Physiol (Lond) 394:315–330

Gray WR, Olivera BM, Cruz LJ (1988) Peptide toxins from venomous *Conus* snails. Ann Rev Biochem 57:665–700

Guggino SE, Guggino WB, Green N, Sacktor B (1987) Blocking agents of Ca^{2+}-activated K^+ channels in cultured medullary thick ascending limb cells. Am J Physiol 252:C128–C137

Habermann E (1968) Biochemie, Pharmakologie und Toxikologie der Inhaltsstoffe von Hymenopterengiften. Rev Physiol Biochem Exp Pharmacol 60:220–325

Habermann E (1972) Bee and wasp venoms. Science 177:314–322

Habermann E (1977) Neurotoxicity of apamin and MCD peptide upon central application. Naunyn-Schmiedeberg's Arch Pharmacol 300:189–191

Habermann E (1984) Apamin. Pharmacol Ther 25:255–270

Habermann E, Fischer K (1979) Bee venom neurotoxin (apamin): iodine labeling and characterization of binding sites. Eur J Biochem 94:355–364

Habermann E, Horvath E (1980) Localization and effects of apamin after application to the central nervous system. Toxicon 18:549–560

Habermann E, Reiz K-G (1965) Ein neues Verfahren zur Gewinnung der Komponenten von Bienengift, insbesondere des zentralwirksamen Peptids Apamin. Biochem Z 341:451–466

Halliwell JV, Othman IB, Pelchen-Matthews A, Dolly JO (1986) Central action of dendrotoxin: selective reduction of a transient K conductance in hippocampus and binding to localized acceptors. Proc Natl Acad Sci USA 83:493–497

Hamill OP, Marty A, Neher E, Sakmann B, Sigworth FJ (1981) Improved patch-clamp techniques for high-resolution current recording from cells and cell-free membrane patches. Pflügers Arch 391:85–100

Harris JB (1985) Phospholipases in snake venoms and their effects on nerve and muscle. Pharmacol Ther 31:79–102

Harvey AL, Anderson AJ (1985) Dendrotoxins: snake toxins that block potassium channels and facilitate neurotransmitter release. Pharmacol Ther 31:33–55

Harvey AL, Gage PW (1981) Increase of evoked release of acetylcholine at the neuromuscular junction by a fraction from the venom of the eastern green mamba snake (*Dendroaspis angusticeps*). Toxicon 19:373–381

Harvey AL, Karlsson E (1980) Dendrotoxin from the venom of the green mamba, *Dendroaspis angusticeps*. A neurotoxin that enhances acetylcholine release of neuromuscular junctions. Naunyn-Schmiedeberg's Arch Pharmacol 312:1–6

Harvey AL, Karlsson E (1982) Protease inhibitor homologues from mamba venoms: facilitation of acetylcholine release and interactions with prejunctional blocking toxins. Br J Pharmacol 77:153–161

Harvey AL, Anderson AJ, Karlsson E (1984a) Facilitation of transmitter release by neurotoxins from snake venoms. J Physiol (Paris) 79:222–227

Harvey AL, Anderson AJ, Mbugua PM, Karlsson E (1984b) Toxins from mamba venoms that facilitate neuromuscular transmission. J Toxicol Toxin Rev 3:91–137

Hermann A, Erxleben C (1987) Charybdotoxin selectively blocks small Ca-activated K channels in *Aplysia* neurons. J Gen Physiol 90:27–47

Hermann A, Hartung K (1983) Ca^{2+} activated K^+ conductance in molluscan neurones. Cell Calcium 4:387–405

Hille B (ed) (1984) Ionic channels of excitable membranes. Sinauer, Sunderland MA

Hodgkin AL, Huxley AF (1952) A quantitative description of membrane current and its application to conduction and excitation in nerve. J Physiol (Lond) 117:500–544

Hoshi T, Aldrich RW (1988) Voltage-dependent K^+ currents and underlying single K^+ channels in pheochromocytoma cells. J Gen Physiol 91:73–106

Howard BD, Gundersen CB (1980) Effects and mechanisms of polypeptide neurotoxins that act presynaptically. Ann Rev Pharmacol Toxicol 20:307–336

Hugues M, Duval D, Kitabgi P, Lazdunski M, Vincent JP (1982a) Preparation of a pure monoiodo derivative of the bee venom neurotoxin apamin and its binding properties to rat brain synaptosomes. J. Biol Chem 257:2762–2769

Hugues M, Duval D, Schmid H, Kitabgi P, Lazdunski M, Vincent JP (1982b) Specific binding and pharmacological interactions of apamin, the neurotoxin from bee venom, with guinea-pig colon. Life Sci 31:437–443

Hugues M, Romey G, Duval D, Vincent JP, Lazdunski M (1982c) Apamin as a selective blocker of the calcium-dependent potassium channel in neuroblastoma cells: Voltage-clamp and biochemical characterization of the toxin receptor. Proc Natl Acad Sci USA 79:1308–1312

Hugues M, Schmid H, Lazdunski M (1982d) Identification of a protein component of the Ca^{2+}-dependent K^+ channel by affinity labelling with apamin. Biochem Biophys Res Commun 107:1577–1582

Hugues M, Schmid H, Romey G, Duval D, Frelin C, Lazdunski M (1982e) The Ca^{2+}-dependent slow K^+ conductance in cultured rat muscle cells: characterization with apamin. EMBO J 1:1039–1042

Janicki PK, Horvath E, Seibold G, Habermann E (1984) Quantitative autoradiography of [^{125}I]apamin binding sites in the central nervous system. Biomed Biochim Acta 43:1371–1375

Jenkinson DH, Haylett DG, Cook NS (1983) Calcium-activated potassium channels in liver cells. Cell Calcium 4:429–437

Jonas P, Bräu ME, Hermsteiner M, Vogel W (1989) Single-channel recording in myelinated nerve fibres reveals one type of Na channel but different K channels. Proc Natl Acad Sci USA 86:17238–17243

Joubert FJ, Taljaard N (1980) The amino acid sequences of two proteinase inhibitor homologues from *Dendroaspis angusticeps* venom. Hoppe-Seylers Z Physiol Chem 361:661–674

Karlsson E (1979) Chemistry of protein toxins in snake venoms. In: Lee CA (ed) Snake venoms, chap 5. Springer, Berlin Heidelberg New York, pp 159–212 (Handbook of experimental pharmacology, vol 52)

Kawai T, Watanabe M (1986) Blockade of Ca-activated K conductance by apamin in rat sympathetic neurones. Br J Pharmacol 87:225–232

Kitabgi P, Vincent J-P (1981) Neurotensin is a potent inhibitor of guinea-pig colon contractile activity. Eur J Pharmacol 74:311–318

Koppenhöfer E, Schmidt H (1968) Die Wirkung von Skorpiongift auf die Ionenströme des Ranvierschen Schnürrings. I. Die Permeabilitäten P_{Na} und P_K. Pflügers Arch 303:133–149

Lancaster B, Nicoll RA (1987) Properties of two calcium-activated hyperpolarizations in rat hippocampal neurones. J Physiol (Lond) 389:187–203

Lebrun P, Atwater I, Claret M, Malaisse WJ, Herchuelz A (1983) Resistance to apamin of the Ca^{2+}-activated K^+ permeability in pancreatic B-cells. FEBS Lett 161:40–44

Lewis RS, Cahalan MD (1988) Subset-specific expression of potassium channels in developing murine T lymphocytes. Science 239:771–775

Maas AJJ, Den Hertog A (1979) The effect of apamin on the smooth muscle cells of the guinea-pig taenia coli. Eur J Pharmacol 58:151–156

Maas AJJ, Den Hertog A, Ras R, Van den Akker J (1980) The action of apamin on guinea-pig taenia caeci. Eur J Pharmacol 67:265–274

MacKinnon R, Miller C (1988) Mechanism of charybdotoxin block of the high-conductance, Ca^{2+}-activated K^+ channel. J Gen Physiol 91:335–349

MacKinnon R, Reinhart PH, White MM (1988) Charybdotoxin block of *Saker* K^+ channels suggests that different types of K^+ channels share common structural features. Neuron 1:997–1001

Mallart A (1985) Electric current flow inside perineural sheaths of mouse motor nerves. J Physiol (Lond) 368:565–575

Marqueze B, Seagar M, Couraud F (1987) Photoaffinity labeling of the K^+ channel-associated apamin-binding molecule in smooth muscle, liver and heart membranes. Eur J Biochem 169:295–298

Mehraban F, Breeze AL, Dolly JO (1984) Identification by cross-linking of a neuronal acceptor protein for dendrotoxin, a convulsant polypeptide. FEBS Lett 174:116–122

Mehraban F, Black AR, Breeze AL, Green DG, Dolly JO (1985) A functional membranous acceptor for dendrotoxin in rat brain: solubilization of the binding component. Biochem Soc Trans 13:507–508

Miller C (1988) Competition for block of a Ca^{2+}-activated K^+ channel by charybdotoxin and tetraethylammonium. Neuron 1:1003–1006

Miller C, Moczydlowski E, Latorre R, Philipps M (1985) Charybdotoxin, a protein inhibitor of single Ca^{2+}-activated K^+ channels from mammalian skeletal muscle. Nature 313:316–318

Moczydlowski E, Lucchesi K, Ravindran A (1988) An emerging pharmacology of peptide toxins targeted against potassium channels. J Membrane Biol 105:95–111

Mourre C, Schmid-Antomarchi H, Hugues M, Lazdunski M (1984) Autoradiographic localization of apamin-sensitive Ca^{2+}-dependent K^+ channels in rat brain. Eur J Pharmacol 100:135–136

Mourre C, Hugues M, Lazdunski M (1986) Quantitative autoradiographic mapping in rat brain of the receptor of apamin, a polypeptide toxin specific for one class of Ca^{2+}-dependent K^+ channels. Brain Res 382:239–249

Mourre C, Cervera P, Lazdunski M (1987) Autoradiographic analysis in rat brain of the postnatal ontogeny of voltage-dependent Na^+ channels, Ca^{2+}-dependent K^+ channels and slow Ca^{2+} channels identified as receptors for tetrodotoxin, apamin and (−)desmethoxyverapamil. Brain Res 417:21–32

Mourre C, Bidard J-N, Lazdunski M (1988) High affinity receptors for the bee venom MCD peptide. Quantitative autoradiographic localization at different stages of brain development and relationship with MCD neurotoxicity. Brain Res 446:106–112

Nanberg E, Connolly E, Nedergaard J (1985) Presence of Ca^{2+}-dependent K^+ channel in brown adipocytes. Possible role in maintenance of α_1-adrenergic stimulation. Biochim Biophys Acta 844:42–49

Naraghashi T, Shapiro BI, Deguchi T, Scuka M, Wang CM (1972) Effects of scorpion venom on squid axon membranes. Am J Physiol 222:850–857

Oberg SG, Kelly RB (1976) Saturable binding to cell membranes of the presynaptic neurotoxin, β-bungarotoxin. Biochim Biophys Acta 433:662–673

Olivera BM, Gray WR, Zeikus R, McIntosh JM, Varga J, Rivier J, de Santos V, Cruz LJ (1985) Peptide neurotoxins from fish-hunting cone snails. Science 230:1338–1343

Othman IB, Spokes JW, Dolly JO (1982) Preparation of neurotoxic [^{3}H]-β-bungarotoxin: demonstration of saturable binding to brain synapses and its inhibition by toxin I. Eur J Biochem 128:267–276

Othman IB, Wilkin GP, Dolly JO (1983) Synaptic binding sites in brain for ^{3}H-β-bungarotoxin – a specific probe that perturbs transmitter release. Neurochem Int 5:487–496

Pappone PA, Cahalan MD (1987) *Pandinus imperator* scorpion venom blocks voltage-gated potassium channels in nerve fibers. J Neurosci 7:3300–3305

Pappone PA, Lucero MT (1988) *Pandinus imperator* scorpion venom blocks voltage-gated potassium channels in GH_3 cells. J Gen Physiol 91:817–833

Pelchen-Matthews A, Dolly JO (1988) Distribution of acceptors for β-bungarotoxin in the central nervous system of the rat. Brain Res 441:127–138

Pennefather P, Lancaster B, Adams PR, Nicoll RA (1985) Two distinct Ca-dependent K currents in bullfrog sympathetic ganglion cells. Proc Natl Acad Sci USA 82:3040–3044

Penner R, Dreyer F (1986) Two different presynaptic calcium currents in mouse motor nerve terminals. Pflügers Arch 406:190–197

Penner R, Petersen M, Pierau Fr-K, Dreyer F (1986) Dendrotoxin: a selective blocker of a non-inactivating potassium current in guinea-pig dorsal root ganglion neurones. Pflügers Arch 407:365–369

Petersen M, Penner R, Pierau Fr-K, Dreyer F (1986) β-Bungarotoxin inhibits a non-inactivating potassium current in guinea pig dorsal root ganglion neurones. Neurosci Lett 68:141–145

Possani LD, Martin BM, Svendsen IB (1982) The primary structure of noxiustoxin: a K^+ channel blocking peptide, purified from the venom of the scorpion Centruroides noxius Hoffmann. Carlsberg Res Commun 47:285–289

Rehm H, Betz H (1982) Binding of β-bungarotoxin to synaptic membrane fractions of chick brain. J Biol Chem 257:10015–10022

Rehm H, Betz H (1983) Identification by crosslinking of a β-bungarotoxin binding polypeptide in chick brain menbranes. EMBO J:1119–1122

Rehm H, Betz H (1984) Solubilization and characterization of the β-bungarotoxin-binding protein of chick brain membranes. J Biol Chem 259:6865–6869

Rehm H, Lazdunski M (1988a) Purification and subunit structure of a putative K^+-channel protein identified by its binding properties for dendrotoxin I. Proc Natl Acad Sci USA 85:4919–4923

Rehm H, Lazdunski M (1988b) Existence of different populations of the dendrotoxin I binding protein associated with neuronal K^+ channels. Biochem Biophys Res Commun 153:231–240

Rehm H, Bidard J-N, Schweitz H, Lazdunski M (1988) The receptor site for the bee venom mast cell degranulating peptide. Affinity labeling and evidence for a common molecular target for mast cell degranulating peptide and dendrotoxin I, a snake toxin active on K^+ channels. Biochemistry 27:1827–1832

Renaud J-F, Desnuelle C, Schmid-Antomarchi H, Hugues M, Serratrice G, Lazdunski M (1986) Expression of apamin receptor in muscles of patients with myotonic muscular dystrophy. Nature 319:678–680

Ritchie AK (1987) Two distinct calcium-activated potassium currents in a rat anterior pituitary cell line. J Physiol (Lond) 385:591–609

Rogawski MA (1985) The A-current: how ubiquitous a feature of excitable cells is it? Trends Neurosci 83:214–219

Romey G, Lazdunski M (1984) The coexistence in rat muscle cells of two distinct classes of Ca^{2+}-dependent K^+ channels with different pharmacological properties and different physiological functions. Biochem Biophys Res Commun 118:669–674

Romey G, Chicheportiche R, Lazdunski M (1975) Scorpion neurotoxin, a presynaptic toxin which affects both Na^+ and K^+ channels in axons. Biochem Biophys Res Commun 64:115–121

Romey G, Hugues M, Schmid-Antomarchi H, Lazdunski M (1984) Apamin: a specific toxin to study a class of Ca^{2+}-dependent K^+ channels. J Physiol (Paris) 79:259–264

Rosario LM (1985) Differential effects of the K^+ channel blockers apamin and quinine on glucose-induced electrical activity in pancreatic beta-cells from a strain of OB/OB (obese) mice. FEBS Lett 188:302–306

Rowan EG, Harvey AL (1988) Potassium channel blocking actions of β-bungarotoxin and related toxins on mouse and frog motor nerve terminals. Br J Pharmacol 94:839–847

Rudy B (1988) Diversity and ubiquity of K channels. Neuroscience 25:729–749

Schauf CL (1987) Dendrotoxin blocks potassium channels and slows sodium inactivation in Myxicola giant axons. J Pharmacol Exp Ther 241:793–796

Schmid-Antomarchi H, Hugues M, Norman R, Ellory C, Borsotto M, Lazdunski M (1984) Molecular properties of the apamin-binding component of the Ca^{2+}-dependent K^+ channel. Radiation-inactivation, affinity labelling and solubilization. Eur J Biochem 142:1–6

Schmid-Antomarchi H, Renaud J-F, Romey G, Hugues M, Schmid A, Lazdunski M (1985) The all-or-none role of innervation in expression of apamin receptor and of apamin-sensitive Ca^{2+}-activated K^+ channel in mammalian skeletal muscle. Proc Natl Acad Sci USA 82:2188–2191

Schmid-Antomarchi H, Hugues M, Lazdunski M (1986) Properties of the apamin-sensitive Ca^{2+}-activated K$^+$ channel in PC12 pheochromocytoma cells which hyperproduce the apamin receptor. J Biol Chem 261:8633−8637

Schmidt RR, Betz H (1988) The β-bungarotoxin-binding protein from chick brain: binding sites for different neuronal K$^+$ channel ligands co-fractionate upon partial purification. FEBS Lett 240:65−70

Schmidt RR, Betz H, Rehm H (1988) Inhibition of β-bungarotoxin binding to brain membranes by mast cell degranulating peptide, toxin I and ethylene glycol bis (β-aminoethyl ether)−N,N,N',N'-tetraacetic acid. Biochemistry 27:963−967

Schweitz H, Stansfeld CE, Bidard J-N, Fagni L, Maes P, Lazdunski M (1989) Charybdotoxin blocks dendrotoxin-sensitive voltage-activated K$^+$ channels. FEBS Lett 250:519−522

Seagar MJ, Granier C, Couraud F (1984) Interactions of the neurotoxin apamin with a Ca^{2+}-activated K$^+$ channel in primary neuronal cultures. J Biol Chem 259:1491−1495

Seagar MJ, Labbé-Jullié C, Granier C, van Rietschoten J, Couraud F (1985) Photoaffinity labeling of components of the apamin-sensitive K$^+$ channel in neuronal membranes. J Biol Chem 260:3895−3898

Seagar MJ, Labbé-Jullié C, Granier C, Goll A, Glossmann H, van Rietschoten J, Couraud F (1986) Molecular structure of rat brain apamin receptor: differential photoaffinity labelling of putative K$^+$ channel subunits and target size analysis. Biochemistry 25:4051−4057

Seagar MJ, Deprez P, Martin-Moutot N, Couraud F (1987a) Detection and photoaffinity labeling of the Ca^{2+}-activated K$^+$ channel-associated apamin receptor in cultured astrocytes from rat brain. Brain Res 411:226−230

Seagar MJ, Marqueze B, Couraud F (1987b) Solubilization of the apamin receptor associated with a calcium-activated potassium channel from rat brain. J Neurosci 7:565−570

Shuba MF, Vladimirova IA (1980) Effect of apamin on the electrical responses of smooth muscle to adenosine 5'-triphosphate and to non-adrenergic, non-cholinergic nerve stimulation. Neuroscience 5:853−859

Silveira R, Barbeito L, Dajas F (1988) Behavioral and neurochemical effects of intraperitoneally injected dendrotoxin. Toxicon 26:287−292

Sitges M, Possani LD, Bayon A (1986) Noxiustoxin, a shortchain toxin from the Mexican scorpion Centruroides noxius, induces transmitter release by blocking K$^+$ permeability. J Neurosci 6:1570−1574

Smith C, Phillips M, Miller C (1986) Purification of charybdotoxin, a specific inhibitor of the high-conductance Ca^{2+}-activated K$^+$ channel. J Biol Chem 261:14607−14613

Solc CK, Zagotta WN, Aldrich RW (1987) Single-channel and genetic analysis reveal two distinct A-type potassium channels in Drosophila. Science 236:1094−1098

Stanfield PR (1983) Tetraethylammonium ions and the potassium permeability of excitable cells. Rev Physiol Biochem Pharmacol 97:1−67

Stansfeld CE, Feltz A (1988) Dendrotoxin-sensitive K$^+$ channels in dorsal root ganglion cells. Neurosci Lett 93:49−55

Stansfeld CE, Marsh SJ, Halliwell JV, Brown DA (1986) 4-Aminopyridine and dendrotoxin induce repetitive firing in rat visceral sensory neurones by blocking a slowly inactivating outward current. Neurosci Lett 64:299−304

Stansfeld CE, Marsh SJ, Parcej DN, Dolly JO, Brown DA (1987) Mast cell degranulating peptide and dendrotoxin selectively inhibit a fast-activating potassium current and bind to common neuronal proteins. Neurosci 23:893−902

Storm JF (1987) Action potential repolarization and a fast after-hyperpolarization in rat hippocampal pyramidal cells. J Physiol (Lond) 385:733−759

Strydom DJ (1973) Protease inhibitors as snake venom toxins. Nature (New Biol) 243:88−89

Stühmer W, Stocker M, Sakmann B, Seeburg P, Baumann A, Grupe A, Pongs O (1988) Potassium channels expressed from rat bain cDNA have delayed rectifier properties. FEBS Lett 242:199−206

Szente MB, Baranyi A, Woody CD (1988) Intracellular injection of apamin reduces a slow potassium current mediating afterhyperpolarizations and IPSPs in neocortical neurons of cats. Brain Res 461:64–74

Tabti N, Bourret C, Mallart A (1989) Three potassium currents in mouse motor nerve terminals. Pflügers Arch 413:395–400

Tanaka K, Minota S, Kuba K, Koyano K, Abe T (1986) Differential effects of apamin on Ca^{2+}-dependent K^+ currents in bullfrog sympathetic ganglion cells. Neurosci Lett 69:233–238

Tas PWL, Kress HG, Koschel K (1988) Presence of a charybdotoxin sensitive Ca^{2+}-activated K^+ channel in rat glioma C6 cells. Neurosci Lett 94:279–284

Taylor JW, Bidard J-N, Lazdunski M (1984) The characterization of high-affinity binding sites in rat brain for the mast cell-degranulating peptide from the bee venom using the purified monoiodinated peptide. J Biol Chem 259:13957–13967

Timpe LC, Schwarz TL, Tempel BL, Papazian DM, Jan YN, Jan LY (1988) Expression of functional potassium channels from *Shaker* cDNA in *Xenopus* oocytes. Nature 331:143–145

Traoré, Cognard C, Potreau D, Raymond G (1986) The apamin-sensitive potassium current in frog skeletal muscle: its dependence on the extracellular calcium and sensitivity to calcium channel blockers. Pflügers Arch 407:199–203

Tzeng M-C, Hseu MJ, Yang JH, Guillory RJ (1986) Specific binding of three neurotoxins with phospholipase A_2 activity to synaptosomal membrane preparations from the guinea pig brain. J Protein Chem 5:221–228

Valdivia HH, Smith JS, Martin BM, Coronado R, Possani LD (1988) Charybdotoxin and noxiustoxin, two homologous peptide inhibitors of the $K^+(Ca^{2+})$ channel. FEBS Lett 226:280–284

Velluti JC, Caputi A, Macadar O (1987) Limbic epilepsy induced in the rat by dendrotoxin, a polypeptide isolated from the green mamba (*Dendroaspis angusticeps*) venom. Toxicon 25:649–657

Vladimirova AI, Shuba MF (1978) Effect of strychnine, hydrastine and apamin on synaptic transmission in smooth muscle cells. Neurophysiology 10:213–217

Weir SW, Weston AH (1986) Effect of apamin on responses to BRL 34915, nicorandil and other relaxants in the guinea-pig taenia caeci. Br J Pharmacol 88:113–120

Weller U, Bernhardt U, Siemen D, Dreyer F, Vogel W, Habermann E (1985) Electrophysiological and neurobiochemical evidence for the blockade of a potassium channel by dendrotoxin. Naunyn-Schmiedeberg's Arch Pharmacol 330:77–83

Wolff D, Cecchi X, Spalvins A, Canessa M (1988) Charybdotoxin blocks with high affinity the Ca-activated K^+ channel of Hb A and Hb S red cells: individual differences in the number of channels. J Membrane Biol 106:243–252

Wu K, Carlin R, Sachs L, Siekevitz P (1985) Existence of a Ca^{2+}-dependent K^+ channel in synaptic membrane and postsynaptic density fractions isolated from canine cerebral cortex and cerebellum, as determined by apamin binding. Brain Res 360:183–194

Zhang L, Krnjevic K (1987) Apamin depresses selectively the after-hyperpolarization of cat spinal mononeurons. Neurosci Lett 74:58–62

Subject Index